Tactile Mapping

Tactile Mapping

Cartography for People with Visual Impairments

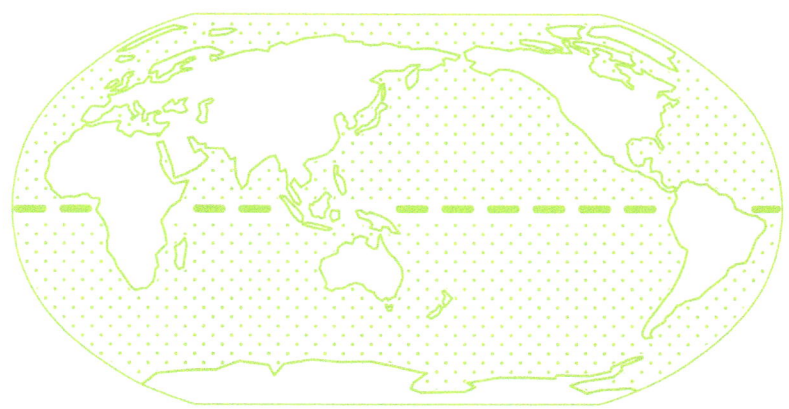

Edited by **Vincent van Altena**
& **Jakub Wabiński**

Esri Press
Redlands, California

Esri Press, 380 New York Street, Redlands, California 92373-8100
Copyright © 2025 Esri
All rights reserved.
Printed in the United States of America.
29 28 27 26 25   1 2 3 4 5 6 7 8 9 10

ISBN: 9781589488359
Library of Congress Control Number: 2025935163

The information contained in this document is the exclusive property of Esri or its licensors. This work is protected under United States copyright law and other international copyright treaties and conventions. No part of this work may be reproduced or transmitted in any form or by any means, electronic or mechanical, including photocopying and recording, or by any information storage or retrieval system, except as expressly permitted in writing by Esri. All requests should be sent to Attention: Director, Contracts and Legal Department, Esri, 380 New York Street, Redlands, California 92373-8100, USA.

The information contained in this document is subject to change without notice.

US Government Restricted/Limited Rights: Any software, documentation, and/or data delivered hereunder is subject to the terms of the License Agreement. The commercial license rights in the License Agreement strictly govern Licensee's use, reproduction, or disclosure of the software, data, and documentation. In no event shall the US Government acquire greater than RESTRICTED/LIMITED RIGHTS. At a minimum, use, duplication, or disclosure by the US Government is subject to restrictions as set forth in FAR §52.227-14 Alternates I, II, and III (DEC 2007); FAR §52.227-19(b) (DEC 2007) and/or FAR §12.211/12.212 (Commercial Technical Data/Computer Software); and DFARS §252.227-7015 (DEC 2011) (Technical Data–Commercial Items) and/or DFARS §227.7202 (Commercial Computer Software and Commercial Computer Software Documentation), as applicable. Contractor/Manufacturer is Esri, 380 New York Street, Redlands, California 92373-8100, USA.

Esri products or services referenced in this publication are trademarks, service marks, or registered marks of Esri in the United States, the European Community, or certain other jurisdictions. To learn more about Esri marks, go to: links.esri.com/EsriProductNamingGuide. Other companies and products or services mentioned herein may be trademarks, service marks, or registered marks of their respective mark owners.

For purchasing and distribution options (both domestic and international), please visit esripress.esri.com.

Contents

Foreword (Jack Dangermond) ... vii

Foreword (Marek Kalbarczyk) ... ix

Part I: Introduction ... 1

Personal story Living without sight (Ellen Zieleman) ... 3

Chapter 1 From visual to tactile: Societal attitudes and accessible information (Vincent van Altena) ... 5

Part II: Maps and perception ... 21

Personal story Defying darkness (Ran Nitka) ... 23

Chapter 2 The relevance of maps in understanding our world (Georg Gartner) ... 25

Chapter 3 Understanding through touch (Astrid M. L. Kappers) ... 41

Case study The development of tactile mapping in Norway (Carl William Lund and Henrik Gulliksen Schüller) ... 59

Case study Dreams become a goal (Ashna Abdulrahman Kareem Zada) ... 63

Part III: Designing tactile maps ... 67

Personal story A journey beyond knowledge (Petr Novák) ... 69

Chapter 4 Map symbol design: Visual and haptic variables (Amy L. Griffin) ... 73

Chapter 5 Map design and cognition (Jakub Wabiński and Simon Ungar) ... 87

Chapter 6 Generalization for tactile maps (Guillaume Touya) ... 105

Case study	Tactile maps of historic gardens (Jakub Wabiński)	117
Case study	Making the invisible visible (Shirly Goldner)	121

Part IV: Users and education   127

Personal story   Completely lost without maps (Leydiane Cristina Santana)   129

Chapter 7   User-centered and inclusive cartographic design (Robert Roth, Merve Keskin, and Zdeněk Stachoň)   131

Chapter 8   Learning geography when you're blind (Carla Cristina Reinaldo Gimenes de Sena and Waldirene Ribeiro do Carmo)   155

Chapter 9   Training in orientation and mobility (Petr Červenka)   171

Case study   A collaborative approach to tactile mapping in the Netherlands (Jolijn Jansen)   189

Case study   Tactile world thematic map (Young-Hoon Kim)   195

Part V: Reliable output   201

Personal story   We drop things at the same rate (Dorothy Atieno Lensa)   203

Chapter 10   Accessible media (Radek Barvíř, Alena Vondráková, and Jan Brus)   205

Chapter 11   Methodical reflections (Albina Mościcka)   225

Case study   3D printed cartography in East Africa (Samuel Foulkes and Quentin Roa)   243

Case study   Optimized route planning for blind pedestrians (Sagi Dalyot and Achituv Cohen)   247

Personal story   Maps aren't just a fun gimmick (Parham Doustdar)   251

Conclusion   (Vincent van Altena and Jakub Wabiński)   253

Personal story   I finally belong (Ellen Zieleman)   259

Acknowledgments   261

Foreword

Jack Dangermond.

One of the frontiers of computer cartography involves making tactile maps for people with visual impairments.

The innovations reported in this book reflect ongoing research at many institutions throughout the world by academics, practitioners, and technicians based on the input, feedback, and needs of people with visual impairments.

It represents the creation of a new type of computer cartography, using standard GIS databases, software, and printers to create tactile, braillelike outputs. The authors not only describe techniques for how to make maps for people with visual impairments but share how the application of these digital outputs can literally change spatial perception for individuals who have never had the opportunity to spatially abstract their world.

My hope is that the information provided in this book will open the application of GIS to a whole new community, broadening access to spatial information for people with special needs.

Jack Dangermond
Cofounder and president of Esri®
Redlands, California
March 2025

Foreword

Marek Kalbarczyk.

As a blind mathematician and IT specialist, a designer of IT systems for visually impaired individuals, a manager, an author of guidebooks and fiction, as well as a sociopolitical activist, I consider the situation of blind individuals in the broadest possible context. I have had the opportunity to present my views at numerous conferences in Poland and abroad, including as a representative of Poland at the 2022 United Nations Convention on the Rights of Persons with Disabilities (CRPD). Here, let me express my heartfelt gratitude for the opportunity to share a few reflections that I believe are deeply important on the modern approach to equalizing life opportunities for blind individuals.

So! In a few words: The world must be for everyone, not just for the privileged few!

For this reason, we are obligated to ensure that the human-constructed world and all its elements designed for all citizens are properly adapted to the needs and capabilities of people with disabilities—including those who are blind. If this is not the case, we fall into the absurdity of promising equal opportunities for all while simultaneously allowing barriers that hinder the lives of our community. In this context, it is essential to do whatever we can to eliminate these obstacles. Such barriers stem both from the very fact of blindness and from the inadequate design and organization of the surrounding environment.

In the modern world, there should be no place for architectural and informational barriers. Infrastructure must be created and improved in such a way that it is accessible to everyone—both for those who use their sight and for those who do not. This leads us to the second major challenge—information accessibility. Consider, for example, the accessibility of architecture, not only as a physical structure but as a set of textual and graphic information describing it (a floor plan, for instance, or a photograph). Without sight, we must still be able to acquire knowledge without relying on vision. This means replacing it with other senses, primarily hearing and touch. The world must "speak" to us and be presented to us in a tactile form.

When every sighted person has effortless access to visual content, which significantly contributes to their development, we must ensure that blind individuals are equally well-informed. We cannot imagine a world where sighted people are illiterate, unfamiliar with the appearance of famous landmarks, works of art, the faces of their loved ones, maps, or terrain layouts. The ease with which sighted people access these elements creates a vast cultural gap between them and us—blind individuals. To bridge this gap, societies must make every effort to provide blind people with a wide range of technological solutions that mitigate these disparities.

Among the most important of these are braille publications, embossed graphics, tactile maps and plans, and 3D models (for the sense of touch), as well as sound-based systems, audio recordings, and audio descriptions (for the sense of hearing). Is this achievable? Technological advancements now enable us to design and produce such solutions, so the answer is a resounding yes! The remaining challenge is securing financial resources. However, if we were to claim that such resources weren't available, our civilization would not deserve to be called modern and humanistic.

I would like to thank the authors, editors, and publishers of this book for their efforts to address these challenges and bring these issues to life. Such initiatives bring us closer to the goal of equalizing life opportunities for blind people. This very publication is the best proof that we, the blind, have every reason to believe that things are moving in the right direction.

Marek Kalbarczyk
President of the Chance Foundation—We Are Together
   (A Chance for the Blind)
Warsaw, Poland
March 2025

Part I
Introduction

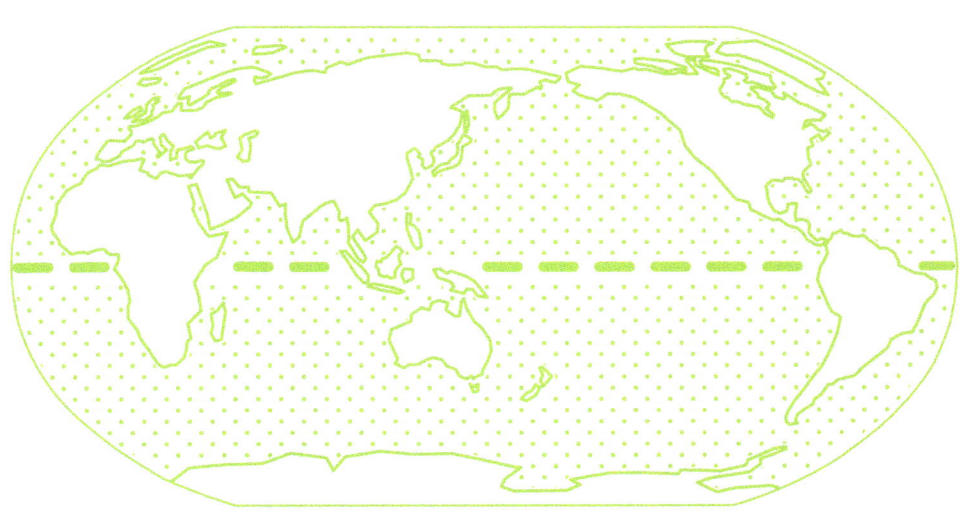

Personal story

Living without sight

Ellen Zieleman

Ellen and her guide dog Sjef.

I was born with a severe visual impairment. Because of persistent and worsening symptoms, I had both my eyes surgically removed a few years ago. Along with my eyes, the last glimmer of light I could see vanished.

You cannot imagine what it's like when your sight disappears forever. You become uncertain and utterly dependent. The confidence and stability I once had disappeared. When I walked, I shuffled, and I had to learn all over again how to navigate with a white support cane. It was an enormous challenge.

Of course, my mapping app on my phone indicates where I need to turn left or right. I also rely on my guide dog, Sjef. However, neither of these resources tells me where it's safest to cross the street or what's around me. That kind of information is hard to come by and needs to be specially tailored for me.

As a result, my spontaneity fades away. I must plan everything in advance, often weeks ahead. And for almost everything, I depend on others. In the best-case scenario, I say to someone, "Take me there." But I often feel discouraged, and instead I'll think, "Never mind."

Chapter 1

From visual to tactile: Societal attitudes and accessible information

Vincent van Altena

It was June 2017 when I crossed paths with Anna Vetter at a conference in Zurich, Switzerland. Our shared passion for maps sparked an animated conversation. Midway through, Anna leaned in and said, "I have something to show you." She disappeared briefly and returned with a black-and-white drawing. As I studied it, the contours of Switzerland emerged; the black lines seemed to leap off the paper.

With contagious enthusiasm, Anna delved into her experiences: her work with geographic information systems (GIS), her internship, and, most intriguingly, her project. It all began when she connected with a blind man who loved maps.

"But how do you make maps accessible to someone who can't see?' I wondered aloud. "Aren't maps inherently visual?" Anna's response surprised me. "Initially, I had the same doubts," she said, "but as I explored further, a whole new world unfolded. Our field has untapped potential to enhance people's geographical understanding. Inspired, I set a goal for my internship: to create a tactile atlas of Switzerland."

I thought, "If Anna can create an atlas for all of Switzerland, surely the Netherlands could be mapped, too." Fueled by curiosity, I asked her if she would be willing to share the project with me. She agreed, promising to email me the details once she returned to the office...

And thus began my journey into the fascinating world of designing maps for the visually challenged. Please join me as I revisit some of my discoveries, insights, and experiences.

—Vincent

Blindness

The language of our senses: How we create our world

Taste, smell, sound, touch, and sight all play their role in informing us about our surroundings. What they register is transported by our nerves and processed by our brain into digestible and actionable information. At this moment, you can read this text thanks to the system of muscles, nerves, and tissue better known as sight—one of our most complex senses.

The actions of opening, closing, and blinking your eyes are controlled by your eyelid muscle. This not only enables you to see but also prevents your eyes from drying. Stability comes from the sclera, the tough white outer layer at the back of your eye, which becomes transparent at the front, forming the cornea. The bending of light rays is initiated at the cornea. At the same time, the amount of light entering the eye is regulated by the iris, the colored part of the eye. This regulation is achieved through the contraction of muscles in the iris that control the size of the pupil, the black center of the eye. This mechanism protects the retina from overexposure and enables vision in low-light conditions.

Light rays are further refracted by the lens behind the iris and focused precisely onto the retina, particularly the macula. This yellow spot is specialized in providing clear central vision and color perception. The retina and macula consist of two types of light-sensitive cells: rods and cones. Low-light conditions are primarily accommodated by the rods, whereas colors and details in bright light are handled by the cones. All this data is converted into electrical signals by the rods and cones and then transmitted to the brain via the optic nerve. The angle of the eyes is adjusted by the eye muscles, enabling the brain to fuse the images into a single, three-dimensional view.

A chain is only as strong as its weakest link

Few of us ponder the mechanisms behind our eyes. We simply expect them to capture the world around us—for instance, the image of the dog in figure 1-1a. However, our vision is a complex interplay of muscles, nerves, tissue, and brain. One malfunction in any part of this system can lead to eye conditions that compromise vision.

A defect in the cones can lead to mild color vision deficiency (also known as color blindness) or even a complete loss of color, where the world is seen in shades of gray. This condition is usually hereditary and is passed through X chromosomes,

a) Normal sight.   b) Hemianopsia.

c) Cataract.   d) Glaucoma.

e) Macular degeneration.   f) Retinitis pigmentosa.

Figure 1-1. Examples of eye conditions. Images courtesy of Roy van Altena.

which is why nearly one in 12 men experience color vision deficiency, whereas only about one in 200 women do. Other causes of color blindness are eye diseases, brain disorders, or medications.

Sometimes, the eyes aren't the issue. Strokes, brain tumors, or head injuries can cause brain damage resulting in hemianopsia, vision loss in either the right or left half of the visual field and affecting one or both eyes (figure 1-1b). Daily activities such as reading, driving, and navigation may be impacted, and early diagnosis and rehabilitation are important to manage the condition.

Aging can also have serious effects on sight. As the lens of the eye ages, its proteins begin to break down and the lens becomes cloudy, forming a cataract. Light cannot always reach the retina properly, resulting in blurred vision with reduced contrast and compromised night vision (figure 1-1c). Surgical treatment can replace the cloudy lens with an artificial lens.

In other cases, multiple factors can cause chronic eye conditions. Glaucoma, a combination of increased eye pressure and damage to the eye nerve, may, for instance, lead to blind spots in the visual field (figure 1-1d). Ultimately, glaucoma can lead to complete blindness, but early detection and treatment prevent vision loss.

Conversely, many older adults are challenged by deteriorating central vision, which impacts the central part of the retina (the macula). This macular degeneration (figure 1-1e) affects the recognition of faces and fine details. Treatment is not always possible, but medicine or special diets may bring some relief.

Incurable and hereditary eye conditions, such as retinitis pigmentosa (figure 1-1f), may gradually affect the light-sensitive cells in the retina: first the rods, leading to night blindness and a reduced, tunneled field of vision, and then the cones, causing blurred vision within the remaining tube of vision. Ultimately, this can lead to severe low vision or blindness.

In addition to these common conditions, visual challenges stem from a multitude of factors, including infections and inflammation, wear and tear, injuries, and other health issues.

Beyond black and white: The spectrum and impact of vision loss

Blindness refers to the complete or near-complete loss of sight that significantly impacts daily life and is difficult or impossible to correct. It's important to understand that blindness exists on a spectrum. Although some individuals may have no vision at all, many retain some degree of residual sight.

This distinction helps explain the apparent discrepancy among global vision loss statistics. The World Health Organization (WHO) reports a broader definition, encompassing both near- and farsightedness, affecting an estimated 2.2 billion people globally.[1] The International Agency for the Prevention of Blindness (IAPB) focuses primarily on distance vision impairment, encompassing blindness and moderate to severe visual impairment. Its estimates indicate 1.1 billion people living with vision loss in 2020.

The causes of vision impairment vary across the globe. General availability of health care, affordability of eye care services, and access to education play a crucial role. Roughly 90 percent of visually impaired individuals reside in low- and middle-income countries.[2] In high-income countries, the leading causes of blindness are age-related macular degeneration, glaucoma, and diabetic retinopathy (an eye disease caused by diabetes that damages retinal blood vessels).

To assess whether an individual has limited vision, several eye tests exist. A common one is the Snellen test, which helps measure visual acuity. If a person has 20/20 vision, they can see details from 20 feet away as clearly as a person with normal vision. If a person has 20/40 vision, they can only see details at 20 feet that a person with normal vision can see at 40 feet. Visual acuity is tested using a standard-sized Snellen chart at 20 feet, lighted to a standard brightness (figure 1-2).

To qualify for public assistance and services, individuals with severe visual impairment are typically examined according to two factors: how well they can see details at a distance (*visual acuity*) and what area a person can see at one time without moving their eyes (*visual field*). In many countries, a person is considered legally blind if their best-corrected visual acuity in their better eye is 20/200 or worse. This means that even with glasses or contact lenses, they can only read the largest letter on a Snellen eye chart from 20 feet away. A person may also be considered legally blind if their visual field is significantly restricted. This condition, often referred to as tunnel vision, limits peripheral vision to 20 degrees or less in the better eye.

When daylight, focus, clarity, or color vanish, life can be daunting. Daily activities, such as reading, writing, driving a car, or recognizing faces, can become difficult or even impossible. Children may experience developmental delays, struggle with education, and face social and emotional difficulties. Adults may have trouble finding employment, experience higher rates of depression and anxiety, and be more prone to social isolation and accidents.

People with visual impairments who actively use rehabilitation tools, such as magnifiers, screen readers, braille, canes, or guide dogs, tend to have better mental

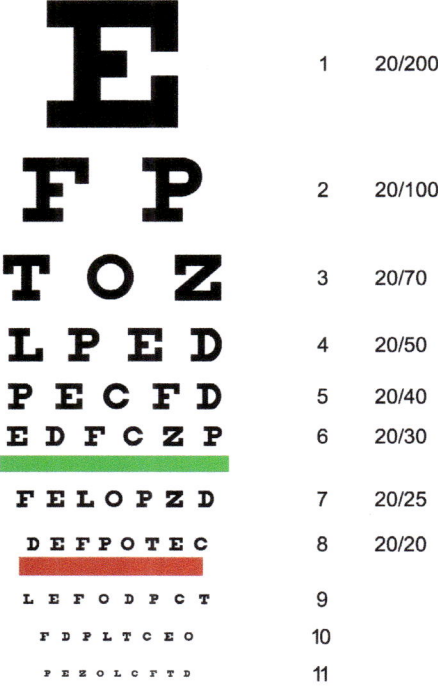

Figure 1-2. A Snellen eye chart used for vision testing. Adobe Stock.

health outcomes and higher levels of social integration. They can navigate their world with confidence and independence, thanks to a combination of specialized training and assistive technologies. Orientation and mobility specialists teach essential skills, such as safe travel, route planning, and obstacle avoidance. Although traditional tools, such as support canes, which often come in white because of their high visibility, and guide dogs, remain invaluable, modern technology has also become an important asset.

Although mobile devices with spoken navigation can provide turn-by-turn directions, they often lack the context and nuance that comes from human interaction. A device can tell you how to get from point A to point B, but it can't convey the subtle nuances of a landscape, changes in elevation, the presence of nearby bodies of water, or the relative position of cities. This lack of a geographic overview not only hinders exploration of unfamiliar neighborhoods but can also impede understanding of local, regional, and international phenomena—for instance, in education or the news. Tactile maps can be a valuable tool to address this challenge.

Historical perspectives

Throughout history, people have sought to provide people with visual impairments with the means to lead fulfilling lives. But historical records are incomplete, and although we see glimpses of these efforts, we can also see that societal attitudes toward blindness have been complex and varied over the centuries.

Some cultures perceived blindness as a divine punishment for moral transgressions. A notable example from the Bible is the disciples' question to Jesus in John 9, inquiring whether the blind beggar's condition was due to his own sins or those of his parents. In contrast, Jesus's response and the examples of various figures in Greek mythology, such as Tiresias, who gained prophetic abilities in exchange for his sight,[3] suggest that blindness could also be seen as a gift, endowing individuals with special insights or roles.

However, throughout history, blindness has often been associated with limitations in contributing to society. This was partly due to the challenges faced by people with visual impairments in learning and understanding, as well as the prevailing view that knowledge acquisition was primarily dependent on sensory experiences, particularly sight. John Locke (1632–1704) argued that knowledge is constructed through sensory experiences and that the absence of sight could therefore limit cognitive abilities. Pushing this idea to extremes, some concluded that education of people with visual impairments was infeasible and a waste of effort. Such views led to social exclusion and marginalization, with blind people being relegated to charity or forced into begging.

Philosophers such as George Berkeley (1685–1753) developed a more nuanced view, suggesting that knowledge and understanding are not dependent solely on sensory experiences but on mental processes as well. This perspective opened possibilities for the education and empowerment of people with visual impairments. Social reformers, inspired by such ideas, advocated practical support, such as the establishment of specialized educational institutions.

From ridicule to revolution: A new era for the blind

In 1771, a French professor, Valentin Haüy (1745–1822), was deeply affected by a performance of nine blind men who entertained a crowd at a café. These men were dressed in absurd costumes, playing discordant music while a "conductor" in wooden clogs and an ass's ears attempted to keep time. Although the crowd found the bizarre performance hilarious, Haüy did not. He was also inspired by his contact with Maria Theresia von Paradis (1759–1824), a blind Austrian pianist who was nicknamed the "Blind Magician." She was a remarkable musician who captivated audiences across Europe.

While touring Germany, Maria Theresia encountered Johann-Ludwig Weissenburg (1752–1800), a young man who had lost his sight to smallpox. Thanks to his teacher, Weissenburg learned to read embossed letters, geometric figures, and maps with raised surfaces and textured materials, which represented features such as water with glass and countries with a variety of grain. During their several meetings, Johann-Ludwig introduced Maria Theresia to these palpable media. Interestingly, these innovations were reserved for the happy few. Weissenburg, for example, was reluctant to make these tools available for the common good, although he shared them with his peers, including Paradis. He wrote to her, "People feel sorry for us, because we cannot see. Friend, do we truly not see?"[4]

Von Paradis's personality and career helped crystallize Haüy's conviction that people with visual impairments could and should be educated. After his experience with the costumed blind men in the public square, he took in the young blind François Le Sueur and taught him to read by touching large carved wooden letters. Appreciating that the blind could read with their fingers, he enhanced his raised alphabet system, which he, as a calligrapher, wanted to look attractive as well as functional for the blind. In 1784, Haüy used these means to teach 12 blind pupils in his school. His methods would become the standard in blind education for the next 50 years.

A tactile revolution: Braille's transformative power

More developments came with Louis Braille (1809–1852), who at a very young age injured his eye with an awl in his father's saddlery shop. Because of inadequate care, an inflammation also affected the other healthy eye, leaving Louis completely blind at the age of three. As a 10-year-old, he entered Haüy's School for the Blind in Paris, where a hundred students had to learn reading from only 14 heavy books that featured tactile but difficult-to-recognize letters.

In the same period, Charles Barbier (1767–1841), a sighted French military officer, had invented a raised-dot system that was based on the 36 sounds of the French alphabet and allowed officers to communicate in the dark. Although the French army never implemented the system, Barbier was convinced that blind students would benefit from it and demonstrated it to Sébastien Guillié (1780–1865), then director of the School for the Blind. Guillié was not very impressed. The young Louis, however, adopted Barbier's system and enhanced it. He simplified the original 12-dot to a 6-dot cell and turned from a phonetic to an alphabetic system. After Braille's simplifications and enhancements, Guillié was willing to give the system a chance.

His successor, Alexandre François-René Pignier (1802–1870), mentor and close friend to Braille, really promoted the use of the new dot script, but the race wasn't over yet. Pignier and his deputy, Pierre-Armand Dufau (1795–1877), struggled for leadership over the institution. When Dufau managed to take over the director's position, Braille's inventive script was deemphasized, and the students were reintroduced to a slightly modernized version of Haüy's embossed type. Despite his efforts, Dufau failed. The blind students simply refused to use the obviously inferior type and continued to use braille script.

All of this did not hinder the young Louis. He turned out to be a gifted student and soon started teaching algebra, grammar, geography, and music at the institute, using his braille alphabet to teach blind students. In 1854, two years after his death, the school finally accepted Louis's superior braille for reading and writing. In 1873, the first book was printed in braille. The script was gradually introduced in all Western European countries and has been the standard convention since 1878. This opened a world of knowledge that people with visual impairments can consume autonomously.

Lesser known than his alphabet are Braille's contributions to standardizing musical braille notation and his invention of *deca-point* and the *raphigraphe*. Deca-point script enabled written communication between blind and sighted people. Together with mechanic Pierre-François-Victor Foucault (1797–1871), Braille developed the raphigraphe (figure 1-3), a needle-writer that transformed the production of deca-point script from cumbersome manual work to highly efficient. In Braille's opinion, deca-point could also be used to communicate about geographic maps. In a brochure published in 1839, he explained how map features from a traditional map (such as borders, mountains, and rivers) could be transferred to palpable media.

A legacy of innovation: Early efforts in tactile technology

Although Braille's contributions are widely celebrated, it's worth noting that others, too, experimented with ways to make materials, such as texts, figures, and maps, accessible to people with visual impairments. One notable example is William Moon (1818–1894), an Englishman who lost his sight at 21 from scarlet fever and would later become a teacher for the blind. His students found reading existing embossed codes difficult, so he developed his own system in 1845, which uses raised curves, angles, and lines. It has been claimed that this type is easier for those who

Figure 1-3. A raphigraphe typewriter for the blind in the Rupriikki Media Museum, Tampere, Finland. Image courtesy of Hotdamnslap.

lose their sight later in life or have a less sensitive touch. In the 1870s, Great Britain introduced the use of braille for writing and Moon type for embossed books in parallel. This failed; by 1886, braille already appeared to be the standard in reading and writing. Moon type has endured and is still in use, although by a small minority.

Interestingly, Moon also experimented with producing tactile maps (figure 1-4), and he certainly was not the only one, as evidenced by 19th- and 20th-century examples of tactile maps using embossed characters and raised lines, as well as woodcuts and wooden 3D globes, all purposed for use in geography education.

What to expect in this book

Fast-forward to 2025. Because of challenges in symbolization and reproduction, along with the need for specialized knowledge, the availability of tactile maps has remained limited. Researchers and practitioners have, however, continued to develop and experiment with tactile symbols and automatic map creation. Recent developments are promising for their standardization, as well as for geography education and mobility training. These efforts are not limited to a specific place, as case studies from different continents demonstrate.

Figure 1-4. Map of England and Wales with tactile elements by William Moon, 1875.
Image courtesy of Perkins School for the Blind Archives.

This book is based on two convictions about cartography. Cartography not only helps us better understand the world but also helps us convey that knowledge to others. However, people with visual impairments encounter challenges every day, which traditional maps are unable to address. Our goal is to provide background, insights, and methods to tailor geographic information for people with visual impairments.

To that end, we've included three types of contributions in this book. Full chapters were written by domain experts to provide an in-depth view on a specific topic. Case studies were added to inspire you to adapt the insights from this book to real life. Finally, personal stories provide valuable insights from the perspectives of people with visual impairments.

In chapter 2 ("The Relevance of Maps in Understanding Our World") in part 2, Georg Gartner explores the role of maps in shaping our perception of the world. Maps transcend mere shapes, lines, and symbols; they serve as gateways for exploring and comprehending our world. From ancient parchment scrolls to digital GIS systems, maps require deciphering. Next, Astrid Kappers explains in chapter 3 ("Understanding Through Touch") how sensory perception through touch works, detailing the cognitive processes involved and highlighting the similarities and differences between haptic and visual perception. She discusses salient features for touch and how the sense of touch is not veridical.

Part 3 focuses on map design. Although color palettes, line thickness, and patterns have traditionally been the building blocks of cartography, a critical question arises: How do you design for someone who cannot see? From this perspective, Amy L. Griffin delves deeper in chapter 4 ("Map Symbol Design: Visual and Haptic Variables") into the subject by exploring and comparing concepts related to visual and haptic variables. She explains cartographic symbol grammars and the capabilities and use of these variables for designing maps for different purposes, such as exploration versus communication and navigation maps versus thematic maps. Next, Jakub Wabiński and Simon Ungar in chapter 5 ("Map Design and Cognition") integrate insights about tactile perception and haptic variables into a chapter on the editing of tactile maps. They discuss the map creation stages, the cognitive "tactualization" approach, and factors influencing, as well as best practices, in tactile map design. Beyond cartographic consistency, content selection is crucial for creating readable and understandable maps. Guillaume Touya in chapter 6 ("Generalization for Tactile Maps") discusses how complex geographic data can be simplified through generalization techniques without sacrificing the essence of intricate landscapes. He provides background on the importance of abstraction, content selection, and multiscale to tactile cartography.

In part 4, Robert Roth, Merve Keskin, and Zdeněk Stachoň explain in chapter 7 ("User-Centered and Inclusive Cartographic Design") how user-centered design focuses on creating maps that resonate with people with visual impairments and how user research can optimally meet individual needs. Under the mantra "nothing about us, without us," they stress a map author's obligation to involve end users in each step of the design process. In chapter 8 ("Learning Geography When You're Blind"), Carla Sena and Waldirene do Carmo discuss effective methods for imparting the art of map reading to people with visual impairments. Teaching map-reading skills is illustrated through user experiences from successful training programs. In chapter 9 ("Training in Orientation and Mobility"), Petr Červenka, with the help of Pavel Wiener, delves further into practical insights, discussing specifics of orientation and mobility training and the role of maps in this training.

The final section, part 5, broadens the book's scope to other media. How can new media, such as the web, mobile apps, and 3D printing, aid in communicating geography? In chapter 10 ("Accessible media"), Radek Barvíř, Alena Vondráková, and Jan Brus explore various media and compare digital, printed, and interactive formats for creating tangible maps and geographic models. They also discuss tactile map production techniques. Albina Mościcka concludes in chapter 11 ("Methodical Reflections") with a methodological reflection on the creation and assessment of tactile maps. She highlights the importance of using methods from traditional cartography, education of the blind, and information technologies to achieve high and repeatable quality in tactile maps. Her chapter underscores the need for a methodological approach to support independent verification and ensure the reliability of tactile maps.

We sincerely hope that all contributions will provide insights that too often remain hidden in scholarly articles and that the case studies from all over the globe might inspire you to try this yourself. Most deeply, we hope that the real-life stories will resonate with you, and that the stories of Ellen, Ran, Petr, Leydiane, Dorothy, and Parham will inspire you!

Further reading

More information on *The Tactile Atlas of Switzerland* is available at www.nationalgeographic.com/culture/article/new-tactile-map-of-swiss-alps-for-the-blind.

Data on the prevalence and global distribution of blindness can be found in World Health Organization, *World Report on Vision* (Geneva: World Health Organization, 2019), https://iris.who.int/handle/10665/328717 and at www.who.int/news-room/fact-sheets/detail/blindness-and-visual-impairment. For an interactive geographic dashboard with statistics on blindness and low vision, visit www.iapb.org/learn/vision-atlas/magnitude-and-projections/global/. The research behind the IAPB atlas is available in "Trends in Prevalence of Blindness and Distance and Near Vision Impairment over 30 Years: An Analysis for the Global Burden of Disease Study," by Rupert Bourne, Jaimie D. Steinmetz, Seth Flaxman, Paul Svitil Briant, Hugh R. Taylor, Serge Resnikoff, Robert James Casson et al., *The Lancet Global Health* 9, no. 2 (February 1, 2021): e130–43.

Numerous sources delve into historical perceptions of the blind, societal attitudes toward them, and efforts to improve their lives. A starting point is the *Encyclopaedia Britannica*, which offers articles on blindness, visual impairment, and the pioneering work of Valentin Haüy. More in depth is Zina Weygand, *The Blind in French Society from the Middle Ages to the Century of Louis Braille* (Stanford University Press, 2009). For a biography of Louis Braille, see C. Michael Mellor, *Louis Braille: A Touch of Genius* (Boston: National Braille Press, 2006).

The Perkins Tactile Maps Collection offers a valuable historical perspective on the evolution of tactile mapping, showcasing a diverse range of tactile maps, dating from the 1830s. These maps, designed for people with visual impairments, depict various geographic features, such as countries, cities, and buildings. Some maps are specifically tailored for the blind, whereas others are adapted from commercial products to include tactile elements: https://flic.kr/s/aHskkmGwvC.

Acknowledgments

Special thanks go to my neighbor, Mr. Jan Kiewiet, a retired general practitioner, for his invaluable input, generously sharing his medical expertise. His willingness to provide feedback on the physiology, pathology, and anatomy sections has been invaluable.

About the author

Vincent van Altena (senior researcher, Kadaster, the Netherlands, co-chair ICA Working Group on Inclusive Cartography) holds a bachelor's degree in theology, an MSc in geographical information science, and a PhD in spatial-temporal interpretation of early Christian literature. At Kadaster, Vincent has worked on topographic mapping, automated generalization, and tailored customer solutions. He participated in international projects, including the European Location Framework, and has also chaired Esri's User Community for Geospatial Authorities Working Group on Map Automation and Generalization. Currently, he leads the Dutch initiative on tactile mapping. Vincent likes cooking (not cleaning), plays piano and synthesizer, and can spend hours reharmonizing music (with different levels of success).

Vincent van Altena.

Notes

1. World Health Organization, *World Report on Vision* (Geneva: World Health Organization, 2019), 26, https://iris.who.int/handle/10665/328717; Rupert R. A. Bourne et al., "Magnitude, Temporal Trends, and Projections of the Global Prevalence of Blindness and Distance and Near Vision Impairment: A Systematic Review and Meta-Analysis," *The Lancet Global Health* 5, no. 9 (September 2017): e888–97.
2. The International Agency for the Prevention of Blindness, "Inequality in Vision Loss," accessed January 8, 2025, www.iapb.org/learn/vision-atlas/inequality-in-vision-loss/.
3. Pindar, *Olympian Ode* 2.84–85; Ovid, *Metamorphoses* 3.317–31.

4   Weygand, Zina, *The Blind in French Society: From the Middle Ages to the Century of Louis Braille*, trans. Emily-Jane Cohen (2003; repr., Stanford University Press, 2009), 77, cf. "Man bedauert uns, das wir nicht sehen, / Freudin, sehen wir wirklich nicht?" Weissenburg in Adolf Kistner, "Christian Niesen, der este Blindenlehrer und sein Schüler Johann Ludwig Weissenburg in Mannheim," *Mannheimer Geschichtsblätter* (1921:12), 205.

Part II
Maps and perception

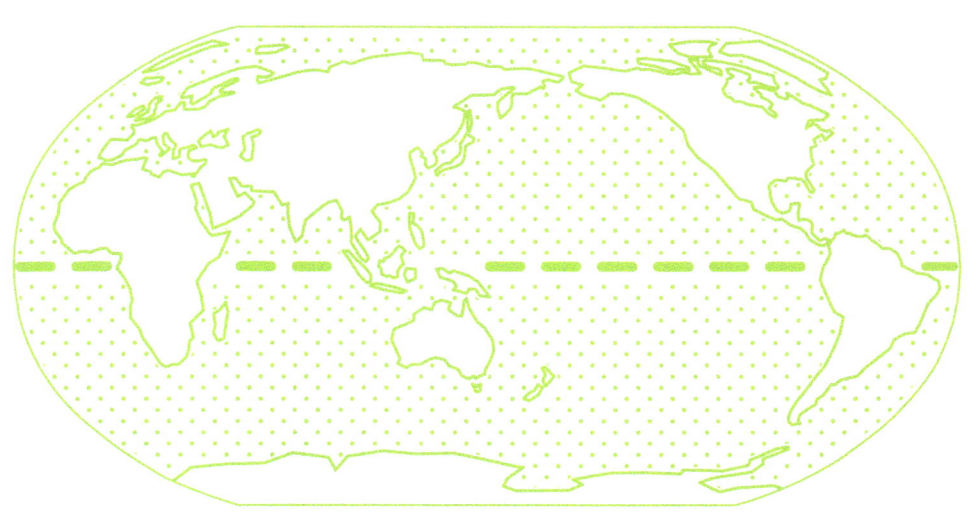

Personal story

Defying darkness

Ran Nitka

Ran in action as a professional balloon artist, in front of his award-winning design.

My challenges with vision began at the age of four, when I experienced early symptoms of visual impairment. By 16, I had retinitis pigmentosa, a genetic disease that would inevitably lead to blindness. The diagnosis was not a shock to me, having witnessed my mother's battle with the same condition. I was aware of the likely course of my life after my mom became completely blind by the age of 40.

My gradual loss of sight culminated in two pivotal moments: on July 21, 2017, I lost vision in my left eye, and on April 24, 2018, I became completely blind in both eyes. These dates mark not only the end of my vision but also the beginning of a profound personal transformation.

At 42, I faced a crossroads. I had two choices: succumb to depression and darkness or embrace life and strive to make a difference. I chose to live life to its fullest and continue my work as a professional balloon artist.

My triumph came during an international balloon conference shortly after losing my sight entirely. Partnering with a sighted colleague, I entered an international balloon competition, relying on a groundbreaking dynamic: My partner provided visual feedback while I planned and crafted balloon sculptures through touch. This innovative collaboration led to a stunning victory. When our first-place win was announced, the audience was astonished to learn that I was blind—a revelation that transformed our team's success into an inspirational story. I came home with a gold medal and the realization that I had achieved what once seemed impossible.

To sustain my career as a balloon artist and motivational speaker, I depend mostly on public transportation. Navigating without sight demands careful preparation, often requiring hours to plan a single trip.

When I was introduced to the Survey of Israel's tactile maps in 2024, I was amazed at how much information I could gather through touch. Being able to feel crosswalks, bus stations, roads, and traffic lights helped me better understand and engage with my surroundings. Having learned braille as a child, I found the maps exceptionally clear and easy to read, further enhancing my ability to navigate independently.

My perceived limitations are my strengths, and now my story extends beyond personal success. At 49, I share my journey through lectures, inspiring others to see adversity not as an obstacle but as a catalyst for growth, creativity, and innovation.

Chapter 2

The relevance of maps in understanding our world

Georg Gartner

I've always felt attracted by maps, and ultimately, this determined my path into cartography. When I was learning to make maps as a student, by hand-drawing contour lines, rivers, and streets and playing around with graphic variables to visualize map objects, I was trained to focus primarily on accuracy and precision. Yet it was experiences like talking to a person with visual impairments, whom I happened to meet at a friend's party, that made me aware of the enormous gap between being technically able to make maps and the real power of maps intended to be user oriented.

Years later, I was involved in research projects related to location-based services, when the possibilities offered by mobile phones led to the development of pedestrian navigation solutions using acoustic and haptic means. When we tested them on children at the Austrian Institute for the Blind, we gained interesting insights into the benefits of tactile maps and multimedia support for wayfinding.

As a cartographer, I believe maps are instrumental for humankind: They allow us to gain awareness, to support decisions, and to communicate spatial information, which is needed in so many aspects of our societies. And I believe that making these instruments equally available to those who are visually impaired is an important achievement.

—Georg

Introduction

Maps are profound representations of knowledge, perception, and understanding. Throughout history, maps have been essential tools for exploration, communication, scientific inquiry, and political organization. For people with visual impairments, maps also serve as essential tools for developing spatial awareness and enhancing independence, enabling them to understand and navigate the world in ways that foster inclusion and empowerment. This chapter explores the multifaceted roles of maps, emphasizing their influence on shaping world views, driving technological advances, and serving as gateways for comprehending our world.

Historical relevance of maps: A journey through time

The history of maps illustrates humanity's evolving perception of the world, from practical tools for survival to reflections of cultural, religious, and scientific priorities. Ancient maps focused on navigation and resource management, such as Babylonian clay tablets for trade (figure 2-1) and Egyptian maps for agriculture. During the medieval period in Europe, maps symbolized spiritual and cultural ideas, while the Renaissance emphasized exploration and empirical accuracy, showcasing humanity's growing curiosity. In modern times, technological advancements, such as satellite observations and digital platforms, have revolutionized mapmaking, making it more precise, accessible, and participatory, reflecting society's enduring drive to comprehend the world.

Maps have always been powerful tools for humanity, serving various functions that have evolved with society's changing needs. From the earliest days, maps were practical instruments, helping humans make sense of the world around them. In ancient times, maps were crucial for societies' survival. Early Babylonians used maps on clay tablets to document trade routes and settlement locations, which allowed them to navigate their world more efficiently. For the Egyptians, maps were essential for agricultural management, especially after the yearly floods of the Nile reshaped the landscape. These early maps were not necessarily accurate in the way we understand today but were immensely valuable for guiding decisions and supporting the needs of their time.

As societies grew more complex, maps began to reflect not just geographic realities but the cultural, political, and spiritual concerns of the people who created them. Ancient Greek maps, for instance, introduced conceptual advances by

Figure 2-1. Babylonian map of the world. Creative Commons.

incorporating ideas such as a spherical Earth, offering a broader understanding of the world's shape. This conceptual shift enabled a more systematic and theoretical approach to representing space, laying the groundwork for future scientific cartography. Maps also played a critical role in empires such as Rome (figure 2-2), where they were used to control and govern vast territories. The Romans relied on maps to illustrate networks of roads, cities, and military positions, making them indispensable tools for administration and the maintenance of power.[1]

During the medieval period (the fifth to 15th centuries), the function of maps shifted to encompass religious and spiritual ideas, representing more than just physical spaces. Maps such as the *Hereford Mappa Mundi* (figure 2-3) were not designed for navigation but instead offered a visual interpretation of a spiritual cosmos, placing religious and cultural significance at the forefront. They mirrored the medieval mindset, focusing on religious understanding rather than empirical geography. This period demonstrated that maps could serve as cultural documents, conveying a society's beliefs and values rather than mere physical locations.[2]

The Renaissance (the 14th to 17th centuries) brought about a transformation in the way maps functioned, driven by the European desire to explore and accurately

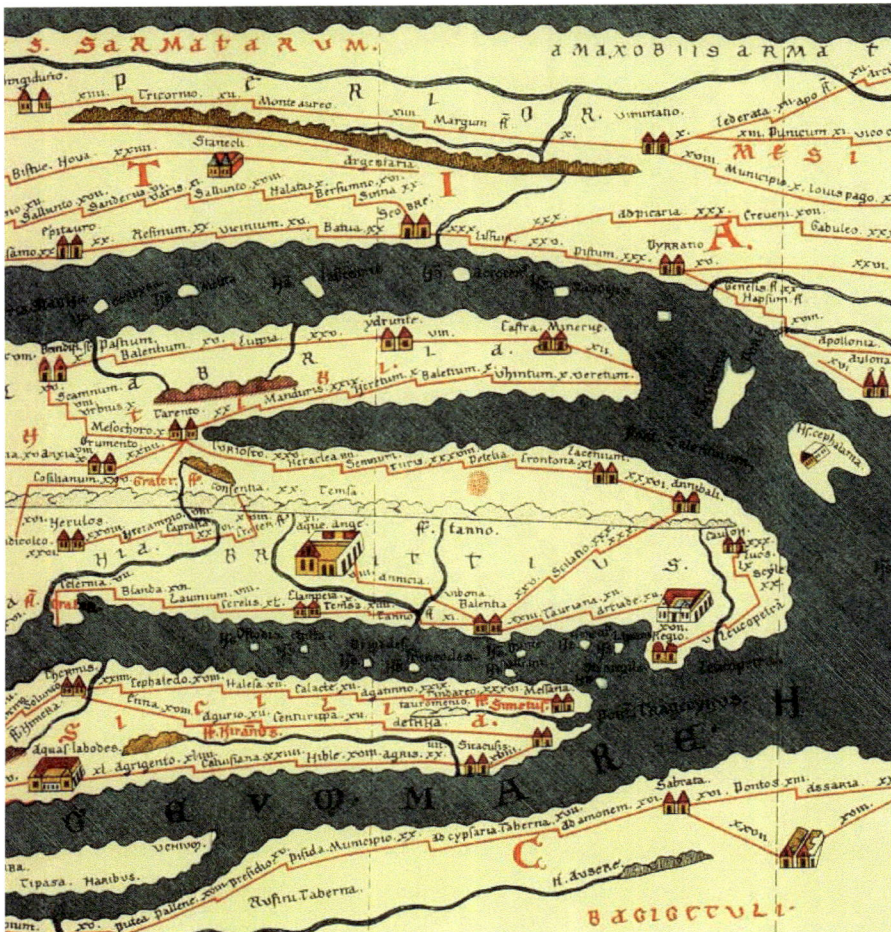

Figure 2-2. Fragment of the *Tabula Peutingeriana*, a well-known reproduction of Roman network cartography. Creative Commons.

depict new lands. Exploration and scientific inquiry demanded maps that could guide long voyages and communicate accurate geographic information. The development of portolan charts, which precisely depicted coastlines and sailing routes, signaled a growing emphasis on empirical observation and precision (figure 2-4). The word *portolan* comes from the Italian *portolano*, meaning "related to ports or harbors." The Renaissance period also saw maps becoming more than practical guides and symbols of wealth and power—they became symbols of discovery and knowledge, reflecting humanity's quest to expand its horizons and understand the world in greater detail.[3]

Figure 2-3. The *Hereford Mappa Mundi*, a 14th-century world map from Hereford Cathedral, England. This map is one of the largest surviving medieval world maps and places Jerusalem at its center, reflecting medieval Christian cosmology. Creative Commons.

In the modern era (16th century to the present), maps have continued to evolve, becoming more sophisticated and essential tools for numerous aspects of human life. The advent of scientific cartography brought a new level of precision to maps, making them indispensable for urban planning, military strategy, and national governance. The use of triangulation and survey techniques allowed for detailed geographic representation, aiding in everything from the development of infrastructure to resource management. Maps became central to decision-making, allowing societies to visualize spatial relationships and make informed choices about land use, resource allocation, and disaster response.[4]

Figure 2-4. Portolan chart by Giacomo Russo of Messina, 1533. The map is hand-drawn on parchment and features intricate details, such as islands, coastlines, and cities marked with flags and symbols. Creative Commons.

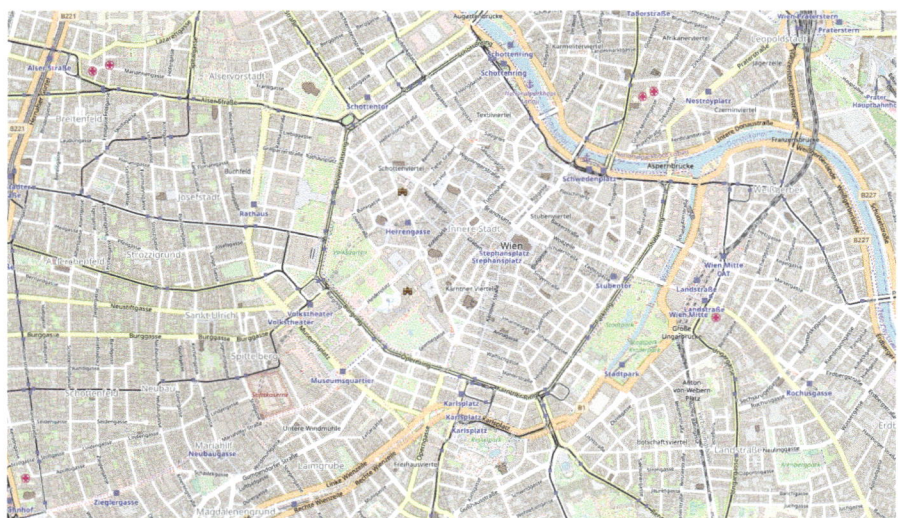

Figure 2-5. Central Vienna mapped by OpenStreetMap. This map was created from OpenStreetMap project data, collected by the community. OpenStreetMap.

The digital age has further expanded the functionality of maps, making them interactive, accessible, and dynamic tools for a range of uses. Modern digital maps serve not only as practical guides but as platforms for exploration, education, and communication. They have become crucial for visualizing complex data, from tracking the spread of diseases to understanding climate patterns. Digital cartography has democratized mapmaking, allowing anyone to contribute to mapping projects and share their perspective. OpenStreetMap is the most prominent example of applying the volunteered geographic information (VGI) principles in cartography (figure 2-5). This shift highlights how maps continue to evolve with technology, adapting to new challenges and opportunities while remaining indispensable tools for understanding and navigating the world. It also marks the postmodernism paradigm of contemporary cartography as a discipline.

Overall, maps have always been more than just representations of physical space; they are reflections of human culture, technology, and our evolving understanding of the world. Whether guiding explorers across uncharted waters, helping governments manage empires, or allowing individuals to navigate urban landscapes, maps have continually adapted to meet the changing needs of society, illustrating how we perceive, organize, and interact with our environment.

Functions of maps

Maps serve a much broader purpose than just guiding individuals from one location to another. They are vital tools that help us visualize, communicate, and make sense of the world's complexities. Whether used in science, education, politics, or culture, maps have a profound impact on how we perceive and interact with our surroundings. The many functions of maps demonstrate their relevance not only to the general population but also to specific groups, such as people with visual impairments, who benefit from accessible maps tailored to their needs.[5]

One of the primary roles of maps is their ability to transform complex data into visual formats that are easier to understand. Maps are indispensable for interpreting large datasets and identifying patterns. This can allow decision-makers and the public to gain awareness, be informed, and make informed decisions. In this light, maps are much more than aesthetic tools; they are critical for understanding trends and making data-driven decisions that affect both local and global communities.

The use of maps as decision-making tools extends to numerous fields beyond urban planning.[6] In disaster management, for instance, maps are essential for coordinating rescue efforts during emergencies. Real-time maps provide a visual overview of affected areas, helping to plan evacuation routes, allocate resources, and deploy rescue teams effectively. Similarly, resource allocation relies heavily on spatial data, with maps guiding logistics, determining optimal locations for new facilities, and managing natural resources. In the military, maps have always been fundamental to strategy and operations. From ancient campaigns to modern warfare, military planners depend on detailed maps to understand terrain, logistical routes, and enemy positions. These examples underscore how maps are not merely representations of space; they are active tools that facilitate complex decision-making across diverse sectors.

In the realm of education, maps play a crucial role in teaching and communication. They are foundational in geography classes, helping students visualize the relationships among countries, regions, and natural features.[7] Beyond geography, historical maps are essential for understanding past events, illustrating the evolution of political boundaries, trade routes, and patterns of exploration. In the sciences, maps serve as tools for communicating research findings, whether showing the distribution of animal species, geologic formations, or ocean currents. These visual summaries help make complex information more accessible, fostering a deeper understanding of the world.

Of course, maps are not neutral representations; they carry political and cultural significance, shaping our perceptions of space and power. National borders on maps are more than just lines; they are symbols of identity, power, and control. The demarcation of territories can assert political claims and reinforce national narratives, often becoming points of contention. This was particularly evident during the colonial era, when European powers employed maps to claim and control vast territories. These colonial maps often ignored Indigenous boundaries, leading to borders that still spark conflicts today. In response to this legacy, there has been a renewed interest in Indigenous mapping traditions, which offer alternative ways of understanding the land. These maps often incorporate spiritual, historical, and ecological knowledge, challenging the dominant Western cartographic norms and providing a richer, more nuanced view of the world.[8]

For people with visual impairments, the significance of maps goes beyond navigation and extends to empowerment and inclusion. Traditional maps, often predominantly visual, exclude those who cannot rely on sight. However, advances in tactile and audio mapping are helping to transform this reality, offering accessible alternatives that make geographic information available to people with visual impairments as well. Tactile maps, with their raised features and braille labels, allow these people to understand spatial relationships and distances. Audio maps provide verbal descriptions of locations, enabling users to navigate and explore independently. These innovations ensure that people with visual impairments are not left out of conversations about geography, history, or science, fostering a sense of independence and inclusion. For cartographers, the ability of these people to engage with maps on equal terms remains a highly important aim.

In addition to promoting independence, accessible maps also play a vital role in the education of people with visual impairments. Geography, which often relies on visual data, becomes more accessible through tactile and audio maps, ensuring that blind students can build an understanding of spatial relationships. These educational tools extend beyond geography to other disciplines, enabling students with visual impairments to explore such topics as history and science in greater depth. Tactile maps can illustrate historical events, showing the changing boundaries of empires, and geologic maps can depict topographical features, giving students a hands-on way to comprehend complex subjects. This inclusive approach to education not only benefits students with visual impairments but also promotes broader awareness of accessibility issues, encouraging educators and institutions to consider diverse learning needs and styles.

As technology continues to evolve, the landscape of accessible mapping is expanding. Innovations in digital maps are providing more interactive and customizable tools to people with visual impairments. Some modern digital maps incorporate haptic feedback, offering vibrations or other tactile sensations when users explore specific features. This multisensory approach enhances the user's spatial experience, making maps more interactive and intuitive. These technological advances are not merely conveniences; they represent a shift toward a more inclusive society where everyone, regardless of visual ability, can access and interpret spatial data.

The cultural and political relevance of maps also extends to people with visual impairments. Accessible maps in public spaces, such as museums or transportation hubs, signal a commitment to inclusivity, ensuring that blind visitors can navigate and enjoy these environments independently. This visibility matters. It emphasizes that accessibility is not a special privilege but a fundamental right. Providing accessible maps is a way to acknowledge the diversity of human experience, recognizing that spatial understanding is not confined to the visual realm. These efforts contribute to a broader cultural shift toward valuing diverse ways of interacting with the world, reinforcing the idea that maps are not just tools for navigation but instruments for inclusion, education, and empowerment.

In recognizing the power of maps to include or exclude, we are reminded that how we map the world reflects how we see it, underscoring the importance of accessible and inclusive cartography.

Technological advancements in mapmaking: A new era of cartography

The technological evolution of mapmaking has profoundly reshaped how we interact with spatial data, turning maps into powerful tools for analysis, prediction, and decision-making.[9] Modern advancements such as GIS, remote sensing technologies, and artificial intelligence (AI) have expanded the capabilities of maps, allowing them to integrate diverse datasets, adapt in real time, and provide actionable insights. These tools emphasize maps as dynamic and data-driven instruments, no longer confined to static representations.[10]

GIS stands at the forefront of this transformation, enabling the combination of spatial and nonspatial data into multilayered visualizations. This integration allows users to analyze complex phenomena, such as predicting flood zones by overlaying topographical and weather data. The real-time capabilities of GIS make it indispensable in contexts such as disaster response or urban traffic management, where dynamic updates ensure relevance and accuracy. Similarly, the inclusion of crowd-sourced data, as seen in platforms such as OpenStreetMap, democratizes mapmaking, filling gaps in coverage and enhancing global accessibility. Beyond GIS, technologies, including satellite imagery and drones, have revolutionized the precision of mapmaking. High-resolution satellite data captures detailed environmental changes, such as deforestation or urban sprawl, while drones provide on-the-ground accuracy, especially in inaccessible regions. 3D laser scanning has opened new possibilities, from creating accurate terrain models to uncovering hidden archaeological sites, showcasing how maps facilitate both scientific discovery and historical preservation.

AI further enhances cartographic capabilities by automating processes and enabling predictive insights. AI-driven models generate real-time maps for disaster response, forecast urban growth for city planning, and predict environmental changes, such as coastal erosion. Also, natural language processing has made digital maps more user-friendly, allowing intuitive searches even with incomplete or colloquial information. These advancements mark a shift from maps as passive tools to active participants in solving real-world challenges. As technology continues to evolve, maps reinforce their role as perceptual tools, not just for navigating space but for shaping how we understand and interact with the world.

Maps as gateways to exploration and understanding

Maps have always served as gateways to understanding, guiding humanity's quest to explore both the physical and social dimensions of the world. As tools of discovery, they not only reveal the unknown but also synthesize complex data into accessible formats, enabling deeper insights into the natural and human landscapes.[11] From uncharted terrains to cultural narratives, maps shape our comprehension of reality and foster exploration in myriad forms. The same is true for tactile maps, which allows independent travel for many people with visual impairments.

In the realm of scientific discovery, maps have been indispensable. They guided polar expeditions, revealing the dynamics of glaciers and previously unknown landmasses, and continue to monitor the Earth's changing poles through satellite imagery, informing climate science. Beyond Earth, maps chart extraterrestrial terrains, guiding space-exploration missions to the moon, Mars, and beyond. Biodiversity mapping has also become crucial for understanding species distribution and prioritizing conservation, providing a vital resource for addressing ecological challenges. Similarly, maps in social and cultural exploration illuminate the diversity of human experience. They chart demographic trends, migration, and socioeconomic disparities, offering a spatial perspective on societal shifts. Linguistic maps showcase the spread of languages and their intersections, whereas heritage and tourism maps capture narratives of cultural significance, reflecting what societies value and how those values evolve over time.

In education and analytic endeavors, maps serve as dynamic tools for learning and research.[12] For researchers, spatial analysis through thematic maps of various kinds uncovers patterns in topics ranging from disease outbreaks to economic activity, offering new ways to visualize complex phenomena. These diverse applications underscore maps as more than just navigation aids—they are vital instruments for understanding the intricate connections within our world, enabling discovery, education, and a deeper appreciation of the spaces we inhabit.

Conclusion

Maps play a critical role in modern societies, serving far beyond their traditional purpose of navigation. They are dynamic tools for analysis, planning, and understanding, with applications ranging from urban development and disaster response to environmental conservation and public health. Technologies such as GIS integrate layers of data—such as population density, infrastructure, and climate—to provide actionable insights for decision-makers. Satellite imagery and remote sensing allow maps to monitor global phenomena, from deforestation and glacier retreat to disease outbreaks, fostering better responses to environmental and humanitarian crises. These capabilities highlight the essential role maps play in addressing contemporary challenges.

For people with visual impairments, maps are equally vital but require adaptation to ensure accessibility. Tactile maps and audio navigation systems empower people with visual impairments to engage with spatial data, supporting independence and inclusion. Technologies such as haptic feedback and voice-guided systems have revolutionized map accessibility, enabling users to navigate urban environments, access transportation networks, or explore geographic knowledge without relying on sight. These innovations underscore the broader societal role of maps: to make knowledge about physical and cultural landscapes accessible to everyone, promoting equality and understanding. In this way, maps remain indispensable tools for navigating not just spaces but also the interconnected challenges of modern life.

Further reading

On the history of cartography

Harley, John B. "The Map and the Development of the History of Cartography." In *The History of Cartography*, vol. 1. The University of Chicago Press, 2018.

Kent, Alexander, and Peter Vujakovic. "Maps and Identity." In *The Routledge Handbook of Mapping and Cartography*. 2017.

Kraak, Menno-Jan, and Sara Irina Fabrikant. "Of Maps, Cartography and the Geography of the International Cartographic Association." *International Journal of Cartography* 3 (sup1): 9–31, 2017.

Kretschmer, Ingrid, et al. *Lexikon zur Geschichte der Kartographie*, 2 vols. Vienna, 1986.

Schiller, Julia. "How Maps Have Been Used: Towards a Typology of Map Functions." MSc cartography thesis, 2023.

Skupin, André, and Charles de Jongh. "Visualizing the ICA: A Content-Based Approach." In *Proceedings of 22nd International Cartographic Conference* (ICC05). ICA, 2005.

On the functions of maps

Fairbairn, David, Georg Gartner, and Michael P. Peterson. "Epistemological Thoughts on the Success of Maps and the Role of Cartography." *International Journal of Cartography* 7, no. 3 (2021): 317–31.

Fairbairn, David, Georg Gartner, and Michael P. Peterson. "Engagement, Communication, and Context: The Success of the Human-Map Nexus." *International Journal of Cartography* (2023): 1–21.

Gartner, Georg. "About the Quality of Maps." *Cartographic Perspectives* 30 (1998): 38–46.

Krzywicka-Blum, Ewa. *Map Functions*. Springer, 2017.

Meng, Liqiu. "Proliferation of Cartographic Education in the Age of Big Data." *Journal of Geodesy and Geoinformation Science* 5, no. 3 (2022): 7–18.

Roth, Robert E., et al. "User Studies in Cartography: Opportunities for Empirical Research on Interactive Maps and Visualizations." *International Journal of Cartography* 3, sup1 (2013): 61-89.

On technological advancements in mapmaking

Cartwright, W., et al. "User Interface Issues for Spatial information Visualization." *Cartography and Geographical Information Systems* (*CaGIS*) 28, no. 1 (2001): 45–60.

Robinson, Anthony C., et al. "Geospatial Big Data and Cartography: Research Challenges and Opportunities for Making Maps that Matter." *International Journal of Cartography* 3, sup1 (2017): 32–60.

On maps as gateways to exploration and understanding

Ceballos, Cantú José Pablo, et al. "Understanding Relevance in Maps Through the Use of Knowledge Graphs." *International Journal of Cartography* 9, no. 3 (2023): 466–87.

Gartner, Georg. "Underpinning Aspects of Developing a Cartographic Curriculum." *Journal of Geodesy and Geoinformation Science* 5, no. 3 (2022): 41–50.

Lambert, Nicolas, and Christine Zanin. *Practical Handbook of Thematic Cartography: Principles, Methods, and Applications*, 1st ed. CRC Press, 2020.

About the author

Georg Gartner is a full professor of cartography at the Vienna University of Technology and head of the Research Unit Cartography. He is currently president of the International Cartographic Association (ICA). He was dean for academic affairs for geodesy and geoinformation at Vienna University of Technology. He is also the organizer of the International Symposia on Location Based Services and the EuroCarto Conference Series. He is editor of the book series Lecture Notes on Geoinformation and Cartography, published by Springer, and editor of the *Journal on Location-Based Services*, published by Taylor & Francis. He serves as president of the Austrian Cartographic Commission, vice president of the Austrian Society of Geodesy and Geoinformation, and is a board member of the Academic Network of United Nations Global Geospatial Information Management. He holds an honorable professorship from Eötvos Loránd University Budapest and the Chinese Academy for Surveying and Mapping. In his free time, he likes skiing, singing, and maps.

Georg Gartner.

Notes

1 Harley, John B., "The Map and the Development of the History of Cartography," in *The History of Cartography*, vol. 1. (The University of Chicago Press, 2018).
2 Kent, Alexander, and Peter Vujakovic, "Maps and Identity," in *The Routledge Handbook of Mapping and Cartography* (2017).
3 Kretschmer, Ingrid, et al., *Lexikon zur Geschichte der Kartographie*, 2 vols. (Vienna, 1986).
4 Schiller, Julia, "How Maps Have Been Used: Towards a Typology of Map Functions," MSc cartography thesis, 2023.
5 Krzywicka-Blum, Eva, *Map Functions* (Springer, 2017).
6 Roth, Robert E., et al., "User Studies in Cartography: Opportunities for Empirical Research on Interactive Maps and Visualizations," *International Journal of Cartography* 3, sup1 (2013): 61–89.
7 Meng, Liqiu, "Proliferation of Cartographic Education in the Age of Big Data," *Journal of Geodesy and Geoinformation Science* 5, no. 3 (2022): 7–18.

8 Fairbairn, David, Georg Gartner, and Michael Peterson, "Epistemological Thoughts on the Success of Maps and the Role of Cartography," *International Journal of Cartography* 7, no. 3 (2021): 317–31.

9 Robinson, Anthony C., et al., "Geospatial Big Data and Cartography: Research Challenges and Opportunities for Making Maps that Matter," *International Journal of Cartography* 3, sup1 (2017): 32–60.

10 Cartwright, William, et al., "User Interface Issues for Spatial Information Visualization." *Cartography and Geographical Information Systems (CaGIS)* 28, no. 1 (2001): 45–60.

11 Ceballos, Cantú José Pablo, et al., "Understanding Relevance in Maps Through the Use of Knowledge Graphs," *International Journal of Cartography* 9, no. 3 (2023): 466–87.

12 Gartner, Georg, "Underpinning Aspects of Developing a Cartographic Curriculum," *Journal of Geodesy and Geoinformation Science* 5, no. 3 (2022): 41–50.

Chapter 3

Understanding through touch

Astrid M. L. Kappers

In the early 1990s, I began research on tactual perception. Compared with the extensive body of work on visual perception, there were relatively few publications in this emerging field, making it feasible to read nearly all the available literature. One day, I encountered a paper by Lederman and Klatzky et al.,[1] pioneers in this area of research. Their findings were hard to believe: Blindfolded observers often failed to recognize simple tangible drawings by touch, even after several minutes of exploration.

The adage "seeing is believing" now had a new twist: "Feeling is believing." Without access to advanced tools such as swell paper printers (which use paper that "swells" when exposed to heat, producing raised images) or 3D printers, my colleagues and I decided to create our own wireframe "drawings." Each of us produced a drawing at home, and initially we served as our own test subjects. To our surprise, recognizing these simple drawings by touch proved to be exceedingly difficult.

Subsequently, we invited several students and colleagues to explore a wireframe "house" and then draw it (without blindfold) when they felt confident. Of the eight participants, four recognized a house, one identified a house from his own drawing, one drew a kind of rabbit, and two produced drawings that only vaguely resembled the original drawing. These results were both surprising and enlightening, underscoring the intricate nature of touch perception.

This insight marked the start of our own research into the perception of tactile drawings.

—Astrid

Our senses are not veridical

All information from our surroundings comes in through our senses. But how reliable is this information? Most people are familiar with visual illusions and understand that what you think you see is not always what is being shown. Notable examples are the café wall illusion (figure 3-1), the illusory pentagon (figure 3-2), or the rotated paintings illusion (figure 3-3). The café wall illusion is a type of geometric-optical illusion where the arrangement of blue and white squares distorts the perception of horizontal lines. In the illusory pentagon, you see lines forming a pentagon that is not actually there. The rotated paintings illusion makes it hard to believe that the paintings are just rotated versions of one another. In addition to these geometric illusions, there are numerous other types of visual misperceptions related to color, luminance, motion, shape, size, and other properties.

Much less known is that our other senses, particularly our sense of touch, are also prone to deceptive perceptions. In the 17th and 18th centuries, philosophers such as George Berkeley (1685–1753) and Étienne Bonnot de Condillac (1714–1780) were not aware of any tactile illusions; they even argued that our sense of touch calibrates our visual system and provides us with the "truth." Because touch involves direct interaction with the environment, and touch illusions go mostly unnoticed in daily life, this idea was not so far-fetched. However, as the following examples show, the information obtained through touch is also often far from veridical.

Misperceptions of temperature

The 17th-century philosopher John Locke (1632–1704) described an experiment in which one hand is placed in a bowl of cold water and the other hand in a bowl of warm water. If, after some time, both hands are placed in a bowl of lukewarm water, the water will feel cold to the hand that was placed in the warm water before and warm to the other. This is an early description of an aftereffect. In the 1980s, Arnold and colleagues[2] performed this experiment with both children and adults. Of course, both groups experienced this temperature aftereffect, and the adults understood that how they perceived the water temperature was related to the previous exposure to another temperature. The children, on the other hand, thought that somehow the water in the bowl had different temperatures and that, by rotating the bowl, they could change their perceptions.

Another intriguing thermal effect is the thermal grill illusion. If you place your hand on a "grill" consisting of interlaced cool (20°C) and warm (40°C) bars, you will experience a burning sensation, although the individual bars will not feel painful at all.

Figure 3-1. Café wall illusion: The parallel horizontal lines seem sloped because of the positions of the background squares. Image courtesy of Astrid M. L. Kappers.

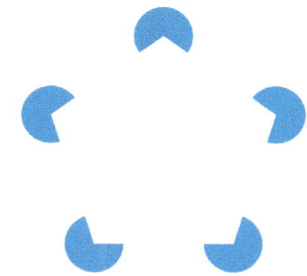

Figure 3-2. Illusory pentagon: There is a vivid perception of a white pentagon, even whiter than the background, but most of the lines of this pentagon are not actually there. Image courtesy of Astrid M. L. Kappers.

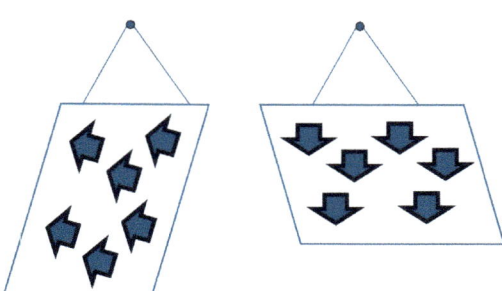

Figure 3-3. Rotated paintings illusion: These paintings are just rotated versions of each other (if you do not believe it, just measure the frames). Image courtesy of Astrid M. L. Kappers.

Curvature and size aftereffects

Almost everything we touch influences the perception of the next touched object. For example, if you hold a small marble in your left hand and a large marble in your right hand for just a short time (less than a minute) and then grasp two identical marbles of intermediate size, the marble in your right hand will feel smaller than the one in your left hand. Curved surfaces also cause aftereffects. If you place your hand on a curved surface, such as a football, and after some time place it on a flat table, the table may feel slightly concave. This effect might not be very strong because you know that the table is flat, but in a well-controlled laboratory setting, such aftereffects can be measured after only two seconds. Interestingly, this effect persists even if you make a fist before touching the flat surface, indicating that the phenomenon involves more than just the cutaneous receptors in the skin; it involves higher-level processing in the brain.

Roughness aftereffects

Rubbing a rough surface with your finger can also alter your perception. After a few seconds, the next surface you touch will feel less rough compared with the perception with another finger. Conversely, if you rub a smooth surface for a while, the next surface you touch will feel rougher. This makes it challenging to objectively estimate roughness, because your perception is influenced by what you previously touched.

Relevance to tactile maps

When comparing textures of different areas on tactile maps, it is useful to remember that it is hard to objectively estimate the roughness and other properties. Although the curvature and temperature examples in this section may not directly relate to the tactile perception of maps, it is important to understand that touch often does not provide accurate information. The following two sections will introduce potential misperceptions that are more directly relevant to the use of tactile maps.

Perceiving drawings through touch

Our first experience with tactile drawings, using the wireframe "house," begged for further investigation. We acquired a swell paper printer, created some simple raised-line drawings, and started informal testing. Blindfolded participants were asked to explore the drawings, and when they felt ready, they were allowed to remove the blindfold and draw what they think they felt. Some representative examples are

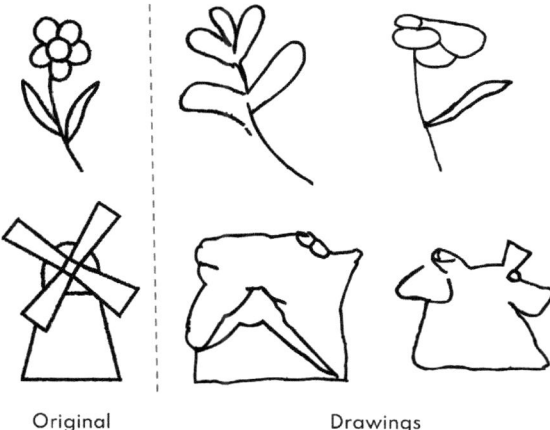

Figure 3-4. Two examples of tactile drawings and how they were perceived. Image courtesy of Astrid M. L. Kappers.

shown in figure 3-4, which show that recognizing drawings by touch is far from easy, even after minutes of exploration.

Given these informal results, we were interested in understanding why it was so challenging to recognize these drawings by touch. We also found it intriguing that some participants who did not recognize the drawing by touch were able to draw what they had felt and subsequently recognized their drawing. Apparently, the necessary information was perceived but not recognized by touch.

Wijntjes and colleagues[3] created 20 simple raised-line drawings of common objects, such as a boat, a duck, an ax, and a tree. The drawings consisted of side views of objects without perspective. Twenty blindfolded sighted participants were asked to explore each drawing for 45 seconds. After this period, they had to guess the object they had felt; they had to provide an answer, even if they were unsure. Following their guess, they had to draw what they had felt, some of the participants with the blindfold still on, others without the blindfold. Regardless of whether they had correctly identified the object, they had to make a drawing. They could take as much time as needed. Upon completing their drawing, they had to guess the object again, with the option to give the same response as before or to change it.

Of the total 400 presented objects, 217 were correctly identified after 45 seconds. More interestingly, of the 91 objects that were initially unidentified and subsequently drawn *without* a blindfold, 31 percent were recognized based on the drawing. As only 2 percent of the unidentified objects were recognized after drawing with a blindfold, this suggests that recognition was not due to extended time or the production of arm movements. Instead, visual information seemed essential to improve the recognition rate.

A follow-up experiment tested whether participants who had to draw blindfolded with a haptic drawing kit (a sketch pad for creating tactile images), and thus received additional haptic instead of visual feedback, also recognized their drawing. This turned out not to be the case, so visual feedback is indeed necessary for the increased recognition rate. In a final experiment, naive observers were asked to recognize the drawings from the first experiment that were identified by those who drew them; these naive observers were also mostly successful. Together, these experiments show that apparently the mental capabilities to recognize raised line drawings are limited; the information to make a reasonable drawing on paper is often there but participants are not always able to integrate this information by mental processing alone.

Comparison to visual recognition

Imagine that you are allowed to look at one of the original raised-line drawings in figure 3-4 for only half a second. Even after this very short time, you will probably be able to produce much more accurate drawings than those shown in figure 3-4. What does this mean for touch compared with vision? One major difference is that in the case of visual inspection, all information—that is, the whole object—is visible at once, whereas in the tactile case, all acquired spatial information must be integrated over time. An illustration of why this process is so challenging is given in figure 3-5: If information is available from only a few isolated spots, it is near impossible to visualize or recognize an object. Of course, this is not the same as tactile exploration of a drawing, but it shows that integration of information is an essential feature of the recognition process. A better illustration would be a circular window moving over this drawing, but that is not feasible in print; when giving such a demonstration in a classroom, it may indeed take several rounds before everyone recognizes the object.

One- versus two-handed exploration

Another aspect that influences the tactile recognition of drawings is the need to locate and trace the lines. During this process, it is common to lose track of the lines. If recognizing a tactile drawing relies on forming a motor image (a mental image of the movement), these tracking errors introduce substantial noise. Magee

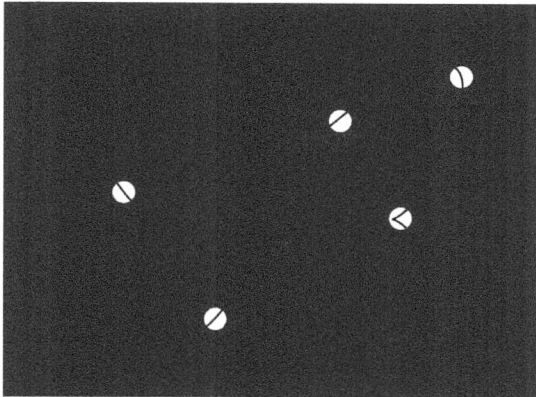

Figure 3-5. A drawing that is mostly covered with a blue screen. The few openings symbolize information that can be gathered with a fingertip. If you want to know what the object is, look at figure 3-11 at the end of this chapter. Image courtesy of Astrid M. L. Kappers.

and Kennedy[4] conducted an experiment in which they guided blindfolded participants' fingers along the lines, resulting in significantly better performance compared with unguided exploration. They also demonstrated that moving these participants' fingers over a pencil drawing, which lacks tactile information, still yielded good performance. This indicates that kinesthetic information (the ability to know where the parts of your body are and how they are moving) is crucial for recognizing tactile drawings.

In an earlier study by Wijntjes and colleagues,[5] hand movements were recorded while participants explored raised-line drawings. Participants were free to choose their own strategy. It was observed that more than 80 percent of the time, participants used both hands, either moving them simultaneously or using one hand as an anchor while the other explored. Wijntjes and colleagues[6] also investigated how performance differed when participants were instructed how to use their hands when exploring raised-line drawings. Participants had to explore half the drawings with one hand and the other half with both hands. The results showed that using both hands significantly improved recognition rates compared with using just one hand.

Relevance to tactile maps

The intention behind exploring tactile drawings and exploring tactile maps is to understand what is depicted. The inherent difficulty of recognizing a raised-line drawing is therefore also applicable to tactile maps. To facilitate the recognition process, the following recommendations are helpful: (1) Because bimanual performance is better than unimanual performance, users could be encouraged to use both hands when using tactile maps; and (2) if a sighted person is also present, this person could be asked to guide the user's finger and, in this way, help build a mental map.

Deformation of visual and haptic space

The ancient Greek architects knew already that to make the facade of a temple like the Parthenon look straight, they had to build it slightly curved. Since the end of the 1800s, scientists have been interested in the so-called curvature of visual space. The famous physician and physicist Hermann von Helmholtz (1821–1894) discovered that when three vertical wires were hung in a straight line directly in front of an observer, the wires were not always perceived to form a flat plane for that observer. Instead, depending on the distance, the "plane" seemed to curve either forward or backward. In the early 2000s, Koenderink and colleagues conducted several intriguing experiments to measure the curvature in more detail. They performed experiments both inside and outside the lab, in more natural circumstances.

One of their studies was an exocentric pointing experiment.[7] Whereas egocentric pointing refers to using your arm as an arrow, exocentric pointing involves directing an arrow at a target, with both arrow and target located at some distance from you. This is like a friend standing some distance from you, pointing toward a bird that she wants you to see. Their equipment included various arrows and targets of different sizes (figures 3-6 and 3-7). The task of the observers was to point the arrow with a remote control at the target. The distances of the arrows and targets were scaled to ensure that all arrows and pointers appeared the same size on the retina. Here and in similar experiments in the lab, participants often "missed" the target in a systematic way, revealing a deformation of visual space. This is one of the reasons that it is often difficult to see the bird that your friend is pointing to.

In 1937, the psychologist Walter Blumenfeld (1882–1967) became interested in the question of whether haptic space is also deformed. He conducted an experiment in which blindfolded observers were asked to hold strings that were pinned with thumbtacks to a table. They were asked to pull gently and to keep these strings parallel to each other, with the strings at varying distances from the observer. He did indeed find systematic deviations from what should be considered physically parallel: When the distance between the wires was small, they diverged toward the observer; for larger distances, they became parallel or even converged.

Inspired by the work on the deformation of visual space and the study by Blumenfeld, we decided to investigate the deformation of haptic space in a more

Figure 3-6. Arrows of different sizes used in the experiment. Image courtesy of Astrid M. L. Kappers.

Figure 3-7. Target spheres of different sizes. Targets were placed on a stand so that the centers of the spheres were always at the same height. Image courtesy of Astrid M. L. Kappers.

Figure 3-8. Picture of an experiment in which a blindfolded participant must rotate a bar such that it points to a target (the disk). The dashed red line shows that the pointing direction is incorrect, missing the target. Image courtesy of Astrid M. L. Kappers.

detailed and systematic way. We designed a series of experiments in which blindfolded participants, sitting behind a table, had to perform several tasks (but in different sessions).[8] The first experiment was a pointing experiment (figure 3-8). Participants had to rotate a bar in such a way that it would point to a target (the disk in the picture). Both target and bar could be placed in several positions on the table. As can be seen in the representative example of figure 3-8, the red line indicating the pointing direction misses the target. The target can be reached only if the line from the bar curves slightly. Other participants missed the target in a similar way, and this was also true for other combinations of target and pointer.

In the second experiment, participants had to rotate two bars in such a way that the bars pointed to each other, or more precisely, became colinear (figure 3-9). The red lines indicate the orientations of the bars, and again there is a mismatch; it is as if the bars can be connected only by a curved line.

In the third experiment, participants had to rotate a bar in such a way that it felt parallel to another bar. In the example of figure 3-10, the task had to be performed with two hands, with the left bar in a fixed orientation. The setting of this particular participant shows a huge mismatch: Instead of parallel, the two bars are almost perpendicular. Not all participants show such large deviations from veridical,[9] but in all cases, the direction of the mismatch is the same. Using just one hand results in similar findings.

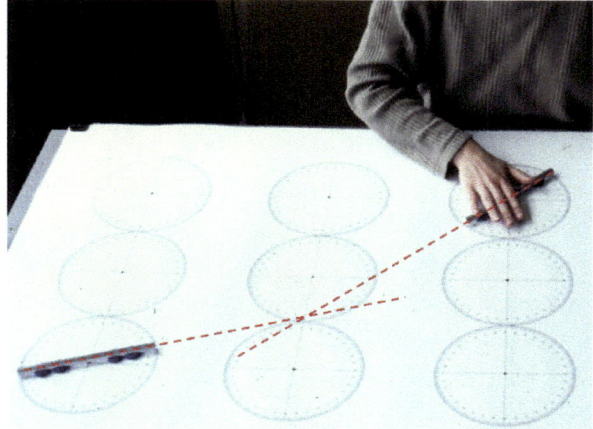

Figure 3-9. Picture of an experiment in which a participant must rotate two bars such that they are aligned—that is, point to each other. The dashed red lines show that the two bars point in different directions. Image courtesy of Astrid M. L. Kappers.

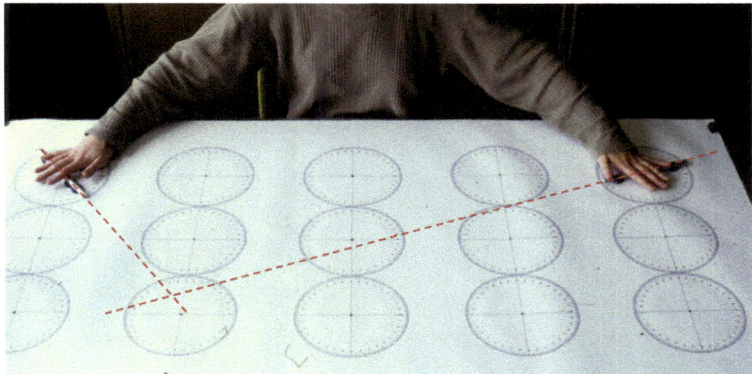

Figure 3-10. Picture of an experiment in which a participant must rotate the right bar in such a way that it feels parallel to the left bar. The dashed red lines show that the two bars clearly have different orientations. Image courtesy of Astrid M. L. Kappers.

How can such deviations be explained? Have a closer look at figure 3-10 and particularly at the hands of the observer. The orientation of the hands with respect to the bars is about the same for both hands. So instead of making the bars parallel to each other, the observer makes the orientation of the bars the same with respect to the hands. That was not the task, but apparently that is how the bars feel parallel to this observer. This would work if the orientations of the arms were the same, but that is clearly not the case. It seems as if perceptually the orientation of the arms is not fully considered. Indeed, for none of the observers did bars that were physically parallel feel parallel. A similar explanation holds for the findings in the other two experiments: If an arm moves from one position to another, the trajectory will be curved, resulting in the typical deviations shown in figures 3-8 and 3-9.

Relevance to tactile maps

These results have direct relevance to the usage of tactile maps, especially the larger ones. Exploring a tactile map involves making a mental image of the layout of streets, buildings, and points of interest. Because of the processes described in the previous section, this mental map might be distorted compared with reality. It seems important to make users aware of this possible distortion and to teach them techniques to minimize the effects, such as attending them to the orientation of their arms when exploring a large map or relating the orientation of streets to the presumably rectangular sides of the map.

Salient features for touch

Many surface properties, such as roughness, structure, and temperature, can be distinguished through touch. Some of these properties can be perceived after just a brief touch and are thus highly salient. In the context of tactile maps, it is useful to have some knowledge of haptic saliency.

Search tasks

The saliency of a property is often investigated through a search task. This is a well-established task in vision research, but this task has more recently also been adapted to haptic research. In an experiment, the observer must decide whether a target with a certain property is present among distractors with slightly different properties, and this must be done as quickly as possible. The idea behind this paradigm is that if a property is salient, the target will "pop out" among the other items (that is, the other objects). In vision, an illustrative example is a red disk among several green disks; for observers who are not color-blind, the red disk will be obvious, and this will not really depend on the number of surrounding green disks. On the other hand, a letter *T* as target among letters *L* as distractors will be much harder to find, and the decision time will clearly depend on the number of distractors. In such a case, one speaks of serial search and the property is not salient.

In touch research, several object and material properties have been investigated in this way. A variety of setups have been used. Sometimes the items are presented on a two-dimensional display, sometimes participants must grasp several three-dimensional items, and sometimes the items are presented to static fingers. Three of the most relevant experiments are described next.

Roughness

Plaisier and colleagues studied the saliency of roughness with circular sandpaper patches on a wooden display.[10] The target, which was either present or not, was a patch with a different roughness than that of the distractors. The number of distractors varied from trial to trial, and their positions also changed. The participants had to sweep their hand over the display and decide as quickly as possible whether the target was present. If the difference between the roughness of the target and distractors was large, participants needed only a single sweep, and they did not even have to touch the target with their fingertips, the most sensitive parts of their hands. However, for small differences, they had to explore all patches one by one before they reached a decision. These results show that, depending on the surroundings, roughness can be a salient feature.

Movability

Ball transfer units were used to study the saliency of the movability of items.[11] In such units, a ball can freely move in its casing. By gluing some of the balls to their casing, the researchers created anchored (immovable) targets and distractors. The units were placed in a two-dimensional irregular grid and the participants had once again to sweep their hand over the display. Their task was to find an anchored target among movable distractors, or vice versa. It was found that a movable target among anchored distractors was easy to find, but an anchored target among movable distractors required visiting the units one by one and thus took much longer. This shows that movability can be a salient feature, but only if the surroundings are stationary.

Temperature

Designing a search experiment involving objects at varying temperatures requires some creativity. The targets used in this case were small brass spheres maintained at a room temperature of about 22°C. Because a target would warm up on contact with the hand, there were multiple copies of the target. The distractors were similar spheres, heated to 38°C by being placed on a plastic layer over a warm-water bath. Thus, the target was cooler and the distractors were warmer than normal hand temperatures. Participants had to grasp several of these spheres and decide whether the target (the cool sphere) was present. It was found that this was an easy task and thus that cool objects pop out among warm objects.

Relevance to tactile maps

Creating a mental image of a tactile map can be facilitated by emphasizing salient features. With printed maps, the options are limited, but using various textures with different roughness will be helpful. Three-dimensional maps offer more opportunities for differentiation. One possibility is to differentiate the height of the various elements; higher ones will be more salient and thus pop out from the background more easily. Although it may not be feasible to heat or cool different areas of a map, using materials such as metal and wood can create a similar perceptual effect. Because of their higher thermal conductivity and heat capacity, metals will feel cooler than wood, even at the same temperature. Additionally, objects that can slightly move will be easier to remember. Lastly, edges and the hardness of objects could be used to make them distinct from other objects.

Conclusion

This chapter introduced several key aspects of touch perception. One important insight is that the perceptual information you obtain from touch often does not have a one-to-one mapping to reality. Everything you touch may influence how you perceive what you touch next. Enhancing tactile maps can be achieved by incorporating features that are particularly noticeable to touch, such as varying textures and differences in temperature or material. Making users aware of potential spatial distortions can be beneficial, and guiding their fingers can help them form a mental map. All these measures together will make using tactile maps by users with a visual disability easier, more useful, and more enjoyable.

Figure 3-11. The drawing that was covered in figure 3-5.

Further reading

On tactile illusions and aftereffects
Hayward, Vincent. "Tactile illusions." *Scholarpedia* 10, no. 3 (2005): 8245.

Kappers, Astrid M. L., and Wouter M. Bergmann Tiest. "Aftereffects in Touch." *Scholarpedia* 10, no. 3 (2015): 32730.

Lederman, Susan J., and Lynette A. Jones. "Tactile and Haptic Illusions." *IEEE Transactions on Haptics* 4, no. 4 (2011): 273–94.

On properties important for touch
Kappers, Astrid M. L., and Wouter M. Bergmann Tiest. "Haptic Saliency." *Scholarpedia* 10, no. 4 (2015): 32734.

On the deformation of visual space
Koenderink, Jan J., Andrea J. van Doorn, and Joseph S. Lappin. "Direct Measurement of the Curvature of Visual Space." *Perception* 29, no. 1 (2000): 69–79.

Acknowledgments

The author thanks Myrthe A. Plaisier of Eindhoven University of Technology, the Netherlands, for her critical reading of the first draft.

About the author

Astrid M. L. Kappers (full professor, Department of Mechanical Engineering, Robotics group, and Department of Industrial Engineering & Innovation Sciences, Human Technology Interaction group, Eindhoven University of Technology [TU/e], the Netherlands) studied experimental physics at Utrecht University, where she graduated in 1985. In 1989, she obtained a PhD at TU/e on the topic of speech recognition. From 1989 to 2012, she worked as assistant, associate, and full professor at Utrecht University. In 2012, she moved with her whole group to the Vrije Universiteit, Amsterdam. Since 2018, she has

Astrid M. L. Kappers.

worked again at TU/e. Her research interests include haptic and visual perception. In 2003, she was the recipient of the prestigious VICI grant. She also participated in several European projects. She was a member of the editorial board of *Acta Psychologica* between 2006 and 2021 and of Current Psychology Letters between 2000 and 2011. She was an associate editor for the IEEE Transactions on Haptics from 2007 to 2011 and from 2017 to 2023. She is also actively involved in the organization of haptic conferences, such as EuroHaptics and WorldHaptics. Her hobbies include bird-watching and photography, reading, cycling, and hiking.

Notes

1. Lederman, Susan J., Roberta L. Klatzky, Cynthia Chataway, and Craig D. Summers, "Visual Mediation and the Haptic Identification of Two-Dimensional Pictures of Common Objects," Perception & Psychophysics 47, no. 1 (1990): 54–64.
2. Arnold, Kevin D., Gerald A. Winer, and Delos D. Wickens, "Veridical and Nonveridical Interpretations to Perceived Temperature Differences by Children and Adults," Bulletin of the Psychonomic Society 20, no. 5 (1982): 237–38.
3. Wijntjes, Maarten W. A., Thijs van Lienen, Ilse M. Verstijnen, and Astrid M. L. Kappers, "Look What I Have Felt: Unidentified Haptic Line Drawings Are Identified After Sketching," Acta Psychologica 128, no. 2 (2008): 255–63.
4. Magee, Lochlan E., and John M. Kennedy, "Exploring Pictures Tactually," Nature 283 (1980): 287–88.
5. Wijntjes, Maarten W. A., Thijs van Lienen, Ilse M. Verstijnen, and Astrid M. L. Kappers, "The Influence of Picture Size on Recognition and Exploratory Behaviour in Raised-Line Drawings," *Perception* 37, no. 4 (2008): 602–14.
6. Wijntjes er al., "Look What I Have Felt," 255.
7. Koenderink, Jan J., Andrea J. van Doorn, Astrid M. L. Kappers, Michelle J. A. Doumen, and James T. Todd, "Exocentric Pointing in Depth," Vision Research 48, no. 5 (2008): 716-23.
8. Kappers, Astrid M. L., and Jan J. Koenderink, "Haptic Perception of Spatial Relations," Perception 28, no. 6 (1999): 781–95.
9. Averaged over many observers, the deviation is about 45 degrees for this situation, instead of more than 60 degrees as shown in figure 3-10.
10. Plaisier, Myrthe A., Wouter M. Bergmann Tiest, and Astrid M. L. Kappers, "Haptic Pop-out in a Hand Sweep," Acta Psychologica 128, no. 2 (2008): 368–77.

11   Van Polanen, Vonne, Wouter M. Bergmann Tiest, and Astrid M. L. Kappers, "Haptic Pop-out of Movable Stimuli," Attention, Perception & Psychophysics 74, no. 1 (2012): 204-15.

Case study

The development of tactile mapping in Norway

Carl William Lund and Henrik Gulliksen Schüller

The work with tactile maps in Norway got started during a user meeting for geospatial authorities in 2023. At this meeting, we, as representatives of the Norwegian Mapping Authority and Geodata AS, were introduced to the concept of tactile maps, which sparked a great deal of engagement. We quickly investigated whether this was something that already existed in Norway and saw that there was limited information on the topic. Therefore, we decided to collaborate and drive this work forward, recognizing the great value of providing more people with access to maps and geographic information.

The first step was to contact the Norwegian Association of the Blind and Partially Sighted (NABP), where our inquiry was well received. Collaborating with NABP proved to be invaluable because it deepened our understanding of the specific challenges faced by people with visual impairments. We then began a process to bring in people with the right expertise in, among other things, cartography, facilitation, and project development to create a plan for a potential project. In Norway, the work has primarily focused on establishing the concept for what a tactile map should include, how it should be designed and distributed, and the user needs and utility values associated with the product. The dialogue so far has highlighted a significant need for tactile maps and demonstrated that they can be a useful tool for many users. Moreover, it has become evident that physical maps can play a crucial role in developing spatial understanding in young people with visual impairments from an early age.

One of the most challenging aspects was putting ourselves in the position of being blind and imagining how tactile maps would be perceived. It was difficult for us to think beyond our training in traditional cartography, where you focus on

Testing of a Tactonom reader during a joint meeting between the parties. Image courtesy of Kartverkett.

using a range of colors, styles, and symbolism to make a map that looks good while still highlighting the important information. To address this, we began by visually analyzing a tactile map to gain an understanding of which symbols and structures seemed easy to distinguish from each other. We then tried to navigate through the same map blindly and quickly found that it was much more complicated to feel the differences than we had initially thought from the visual assessment. Moreover, scale and celestial direction became, to a greater extent, abstract concepts, and we soon realized that we required more expertise to create good readable tactile maps.

NABP's collaboration guided us in removing elements that could be perceived as noise for people with visual impairments, emphasizing the need for simplicity

Draft of a tactile map based on feedback from the Norwegian Association of the Blind and Partially Sighted (NABP). The map includes labels in braille for key landmarks and points of interest and contains a north arrow. Image courtesy of Kartverkett.

and space in the map. Their feedback drove us to focus on creating a minimalist map that could be effectively used as a tactile component. For example, we learned that accessibility was one of the most important elements in the map. In contrast to traditional cartography, which focuses on details to enable navigation based on the real world, a tactile map emphasizes possible accessibility to orient a user. This means in practice that it was not necessarily so significant what type of terrain was present but whether this terrain type was possible to cross without major obstacles. The same applied to adding elements that are not usually incorporated into normal maps, such as stairs. For us, this switch was a lesson in empathy as well as in cartography.

The plan going forward is to establish a formal project to launch and further develop the concept we are working on. The company Geodata AS has begun looking into the idea of an on-demand self-service web solution with the aim of scaling the service to a nationwide level if feasible. The goal is to create a digital solution that allows potential users to extract a section of their area of interest at multiple scales. We also want to enhance accessibility by exploring multisensory experiences, combining tactile maps with audio. One promising solution is the Tactonom reader with maps made from the swell paper technique. This device uses a camera to recognize tactile elements on the map and provides audio descriptions, empowering users to independently explore and understand the information.

Through collaboration and innovation, we can break down barriers and create a more inclusive future. By working together with people with visual impairments and the NABP, we can ensure that everyone has the opportunity to benefit from the power of maps.

Case study

Dreams become a goal

Ashna Abdulrahman Kareem Zada

It's often said that a dream becomes a goal when you take action to achieve it. This is an idea that I like to keep in mind because it focuses on the goal, not the obstacle. As an example, I would like to describe an atlas that I developed for students with visual impairments in Kurdistan. The atlas project, based on the Kurdistan school curriculum, includes 90 tactile maps covering physical, political, and thematic topics for blind and visually impaired Kurdish learners.

Every parent wants their child to have the opportunity to lead a better life than previous generations; I benefited from such an opportunity, studying abroad and earning a PhD in cartography and geoinformatics. However, in my country, Iraq, the lack of tactile maps in the educational system, particularly in parts of Kurdistan, meant that blind and visually impaired students were not instructed in the same curriculum as sighted students. I was determined to use my training in GIS to serve these people with visual impairments.

I am the first person in Iraq to create tactile maps for educational purposes, and this is the first such project in Iraq. In the initial step, I took maps from the school curriculum, from primary to secondary school, re-created them using ArcGIS®, and tested them with visually impaired students at the Runaki Institute in the city of Erbil. The Runaki Institute (Kurdish for "light") was established in 1990 to serve people with visual impairments in the Erbil Governorate (population 1.7 million) and currently serves some 80 to 120 students. This is the only center in the Erbil Governorate that can provide educational services to students with visual impairments, although many other centers have opened recently in the Kurdistan region.

Before creating the atlas, I had conducted a questionnaire among children with visual impairments (ages 6–18 years) at the Runaki Institute to learn their needs and preferences. One of the interesting outcomes was that there was no clear preference among them for either digital or paper maps. I then gathered all the necessary

Tactile Mapping

The author using a swell paper fuser to create tactile maps.

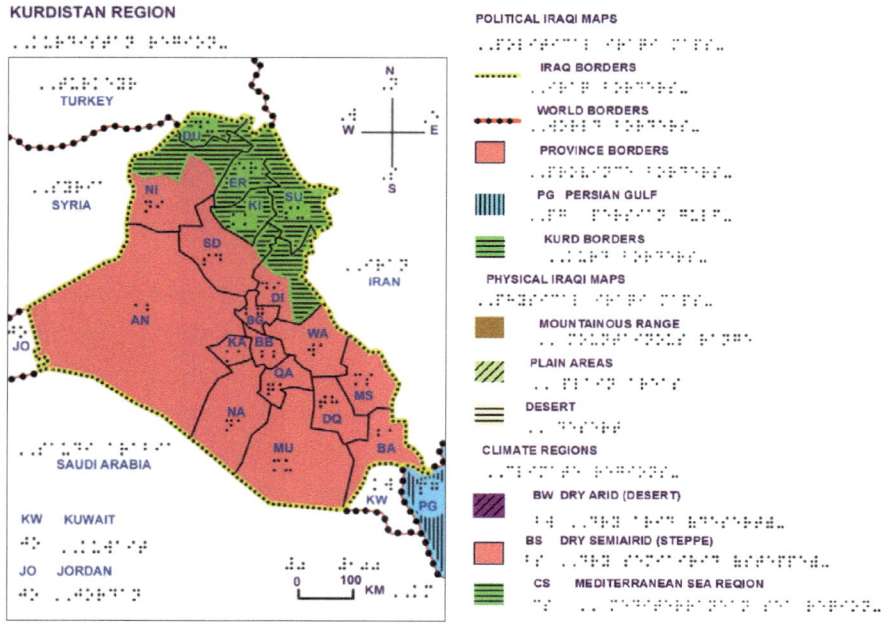

On the left, a digital file of one of the maps developed for the tactile atlas—an overview map of the Kurdistan Region depicting country borders, geographic features, and regions of Iraq. Both large-print and braille labels are applied. On the right, a legend with symbols appearing on a series of maps: political, physical, and climate regions.

data from open sources and, using GIS, created a set of highly contrasting and generalized maps containing graphic and tactile content for both visual and haptic perception.

One of the main design challenges was the use of toponyms. Because the Kurdistan region is inhabited by various nations and tribes, it wasn't possible to include labels in all the languages. Thus, the most reasonable approach was to use English for all the text on maps. In practice, the maps use two-letter abbreviations for most geographic features, explained on separate code sheets.

Finally, some of the developed maps with corresponding legends and code sheets were printed on swell paper and handed to the students. I observed, to my great delight, how happy they were to use these maps in the classroom. Unfortunately, because of a lack of funds and materials, I was able to print only a few examples using the swell paper technique.

Students with visual impairments can use this atlas, with 90 maps, to access the information and knowledge that has been denied them for so long. Tactile maps can help such students learn geographic information while using, maintaining, and improving their abilities and skills. I am proud of the progress of my people in this regard, and I am optimistic. Partly, my optimism is based on seeing the huge wave of support from other countries and from young people around the world who want their governments to invest in this tool for education. They live in a much more globally connected world than the one I grew up in and are among our strongest advocates for learning. My optimism is also based on my confidence that GIS can deliver results and is ready to deliver more and more.

Part III
Designing tactile maps

Personal story

A journey beyond knowledge

Petr Novák

Petr with city miniatures.

I have been visually impaired since birth and completely blind since I was 10 years old. Heeding my lifelong passion for travel and railways, I studied in the Faculty of Transportation Sciences at Czech Technical University in Prague. This is where I discovered the profound impact that tactile maps could have on my understanding of the world.

While I was a student, a close friend painstakingly drew a special map for me. On a sheet of swell paper, this map represented the tram network of Prague, capital city of the Czech Republic. Using this tactile map, along with an electronic timetable, I traced with my fingers the precise locations of hundreds of tram stops and landmarks throughout the city. This made my conceptualization of the city very clear, and the map became my window to Prague, offering me a level of detail that sighted people typically take for granted.

Soon, I added similar tactile maps to my collection, including maps of the Czech cities of Brno and Ostrava and transportation systems, such as the S-Bahn in Berlin, Germany, and the U-Bahn in Vienna, Austria. Although the shapes of these cities and systems remained obscured from my eyesight, their tactile representations opened new horizons for me, revealing each location's intricate web of routes and connections.

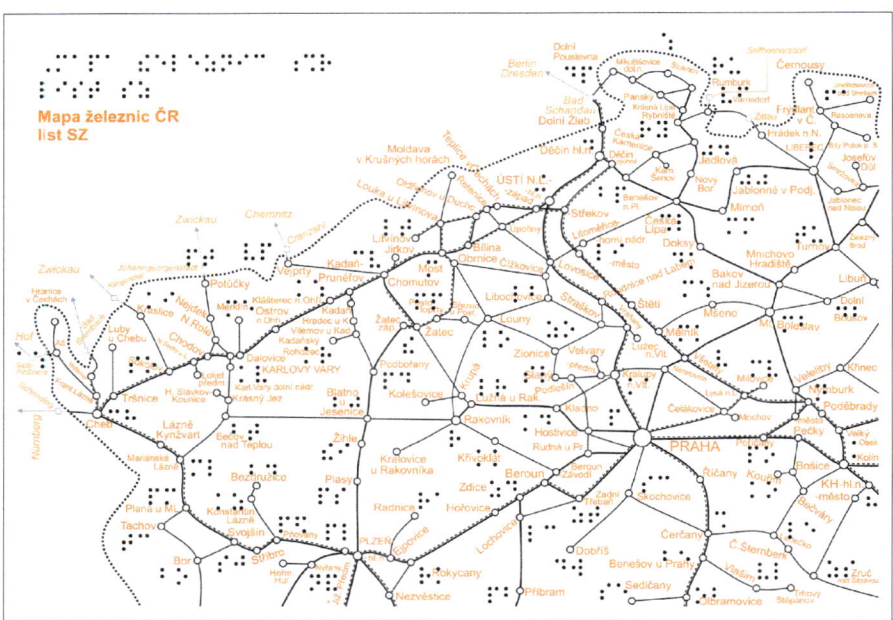

Tactile map of the Czech railway network. Image courtesy of Czech Technical University in Prague.

A real game changer for me was when my university developed a professional-level tactile map of the Czech railway network. This map transformed my knowledge of the Czech Republic, allowing me to pinpoint the locations of thousands of villages, both large and small. Supplementing this tool with an electronic timetable, I embarked on an ambitious quest to travel my home country's entire railway network.

I also traveled through Europe, which yielded intriguing discoveries. Take Narvik, Norway, for instance. Situated above the Arctic Circle, the seaport town of Narvik is home to the world's northernmost standard-gauge railway station. Although standard-gauge tracks enable trains from different regions to travel on the same line, Narvik is connected by rail only to Sweden and not to any other places in Norway—due, in part, to Sweden developing the train line to transport iron ore and, in part, to the rugged Norwegian geography around Narvik. The tactile map I used vividly illustrated this logistical quirk.

The town of Gorizia, Italy, offers another geopolitical peculiarity that I can experience on a tactile map. After World War I, the town—then known as Görz—was incorporated into Italy. But the 1947 Treaty of Paris divided it into two separate entities. The larger part, with the Gorizia Centrále rail station, remained in

Italy, whereas the smaller section, with the Nova Gorica rail station, became part of Yugoslavia (now Slovenia). Despite their proximity, the two stations lack a direct link. Feeling this disconnect on a tactile map gave me a greater understanding of the region.

For blind individuals like me, tactile maps are more than just tools for navigation; they are gateways to independence and exploration. Tactile maps of public transportation grant me the freedom to not only traverse cities and nations but also gain deep appreciation for technology, literature, history, and international relations. Tactile maps can transform blank pages into richly detailed guides that empower people with visual impairments to venture far beyond the limitations of their sight.

Chapter 4

Map symbol design: Visual and haptic variables

Amy L. Griffin

I first became interested in nonvisual data representation when I was an MSc student. That was a time when virtual reality technologies were starting to change what we thought was possible for exploring and communicating geographic information. Interactive haptic displays could be imagined and realized with devices such as the Phantom haptic device, which allows users to feel physical properties of virtual 3D objects. This led me to conduct some research that proposed how different haptic sensations might be used for making maps and how to explore these sensations to build haptic symbols. Fast-forward 25 years, and we now have an even greater range of options for creating touch-sensitive maps, such as 3D printing and interactive, touch-sensitive displays, perhaps now even more widely available than tactile printers, that print on microcapsule papers. To my mind, there is little excuse today for not making maps more accessible to people with vision impairments, a goal I continue to work toward.

—Amy

Introduction

Maps are inherently symbolic. Mapmakers use symbols to represent (or stand for) something from the world. Sometimes these symbols are quite abstract—for example, when a city is represented using a small circle (figure 4-1). Without a map key or legend, this abstract symbol could stand for anything, and the map user may have difficulty knowing what kind of thing is shown on the map. At other times, the symbol may be more figurative, or iconic—the appearance of the symbol may be in some way similar to what it stands for on the map. For example, a tree-shaped point symbol might be understood by the user to represent a forest, even without a map key. Whether a symbol is abstract or iconic, we can deconstruct any given map symbol into a set of even more basic characteristics, such as shape or color. For visual maps, these characteristics are usually referred to as visual variables. For tactile maps, we can identify a corresponding set of haptic variables. These sensory variables, whether they are visual or haptic, are generally considered to be the fundamental building blocks of all map symbols.

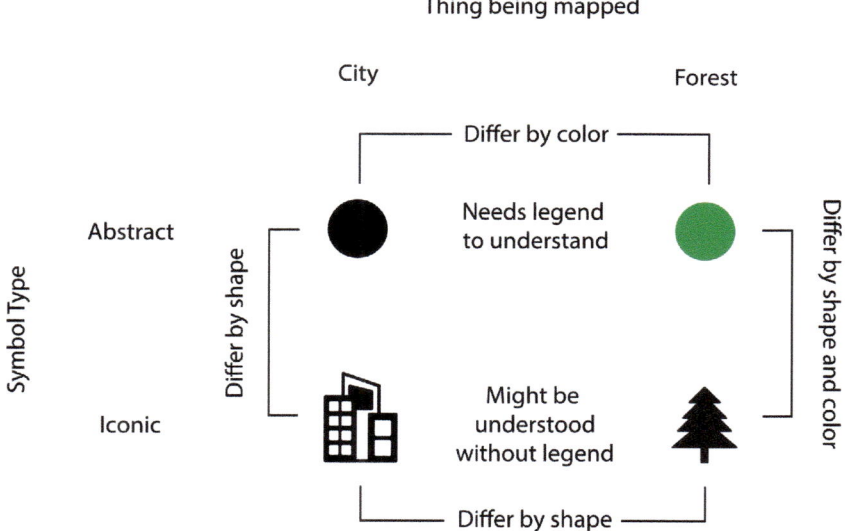

Figure 4-1. Map symbols can be abstract or iconic and are constructed from sensory variables. Image courtesy of Amy L. Griffin.

Cartographic symbol grammars

The first person to write about visual variables was the French researcher Jacques Bertin (1918–2010).[1] He proposed a set of seven graphic (visual) variables combined with a set of rules for how these variables should be used to construct map symbols. These visual variables are a form of cartographic language, and the rules are a cartographic symbol grammar. Bertin's grammar specified whether each visual variable was suitable for representing different types of information: qualitative (nominal), ranked (ordinal), or quantitative (numerical).

Other researchers considered Bertin's list of visual variables to be incomplete and later added more visual variables to his grammar or argued for the use of different terminology to describe a given visual characteristic. Table 4-1 lists visual variables that can be used to form map symbols and provides an assessment of the suitability of each visual variable for representing qualitative, ranked, or numerical information.

The visual phenomenon of color is usually decomposed into three distinct visual variables: hue, lightness, and saturation. *Hue* derives from the predominant wavelength of light that is reflected by the color, also sometimes known as color names, such as red, brown, or yellow. *Lightness* describes the shade of the named color: for instance, light versus dark blue. *Saturation* refers to the intensity or vividness of a hue, with low-saturation colors appearing grayish and high-saturation colors being pure hues.

Shape describes the geometric or iconic form of a symbol. *Arrangement* is the visual variable that describes the spatial distribution of subcomponents of a symbol. For example, for line symbols, two arrangements might be dot-dot-dash versus dot-dash-dot-dash. Symbols with different *orientations* differ from each other by their angle of rotation. Symbols of different *size* might have variations in their length, area, or volume, depending on the symbol's shape. Finally, *texture* is a visual variable that is expressed by the numerosity and density of the symbol's subcomponents, ranging from coarse (few subcomponents, low density) to fine (many subcomponents, high density).

Although most visual maps that are produced use visual variables to create their symbols, researchers have extended the idea of visual variables to other senses, including the sense of touch. The sense of touch is responsible for haptic sensations. Haptic sensations can themselves be differentiated into two types: tactile and kinesthetic sensations. Tactile sensations come from the skin encountering another material, whereas kinesthetic sensations come from gravity's forces on muscles and

Table 4-1. Visual and haptic variable grammars

Variable	Suitability for representing the information type		
	Qualitative	Ranked	Numerical
Visual variables			
Color hue	Good	Only if selected hues are logically ordered	Only if selected hues are logically ordered
Color lightness	Poor	Good	Marginal
Color saturation	Poor	Good	Marginal
Shape	Good	Poor	Poor
Arrangement	Marginal	Poor	Poor
Orientation	Good	Marginal	Marginal
Size	Poor	Good	Good
Texture	Good	Marginal	Marginal
Haptic variables: tactile			
Vibration amplitude	Poor	Good	(not proposed)
Vibration frequency (flutter)	Poor	Good	(not proposed)
Pressure	Poor	Good	(not proposed)
Temperature	Poor	Good	(not proposed)
Haptic variables: kinesthetic			
Resistance	Poor	Good	(not proposed)
Friction	Poor	Good	(not proposed)
Kinesthetic location	Poor	Good	(not proposed)
Haptic variables: tactile analogs of visual variables			
Tangible size	Poor	Good	(not proposed)
Tangible elevation	Poor	Good	(not proposed)
Tangible shape	Good	Poor	(not proposed)
Tangible texture	Good	Marginal	(not proposed)
Tangible orientation	Good	Marginal	(not proposed)
Tangible arrangement	Marginal	Poor	(not proposed)

Note: Marginal ratings mean that this variable may be used effectively in some situations but not others. Adapted from White (2017), Griffin (2001), Ranasinghe and Degbelo (2023).

joints, which change according to where a given body part is located within space. In a haptic variable grammar (table 4-1), we can identify variables that come from tactile sensations (such as vibration or temperature) or from kinesthetic sensations (such as resistance or friction).[2] We can also construct haptic variables that have visual variable analogs. These are distinguished from the visual variables by adding the adjective "tangible."

Several haptic variables do not have a direct analog in the visual variables. The tactile variables of *vibration amplitude* and *vibration frequency* (also called *flutter*) both involve repeated back-and-forth movements around some central point, with vibration amplitude capturing the size of the movement and vibration frequency how often the movement occurs. *Pressure* is a measure of how much the skin is deformed when it is touching something. *Temperature* is the perception of a change in warmth or coolness compared with the skin's temperature. The kinesthetic variable *resistance* is the amount of force needed to deform a surface, whereas *friction* is the stickiness of a surface as the hand moves across it. *Kinesthetic location* is the position of the hand in relation to the body.

These variables can be operationalized in different ways for mapping. Figure 4-2 shows an example of a map that gives its users the option to choose one of

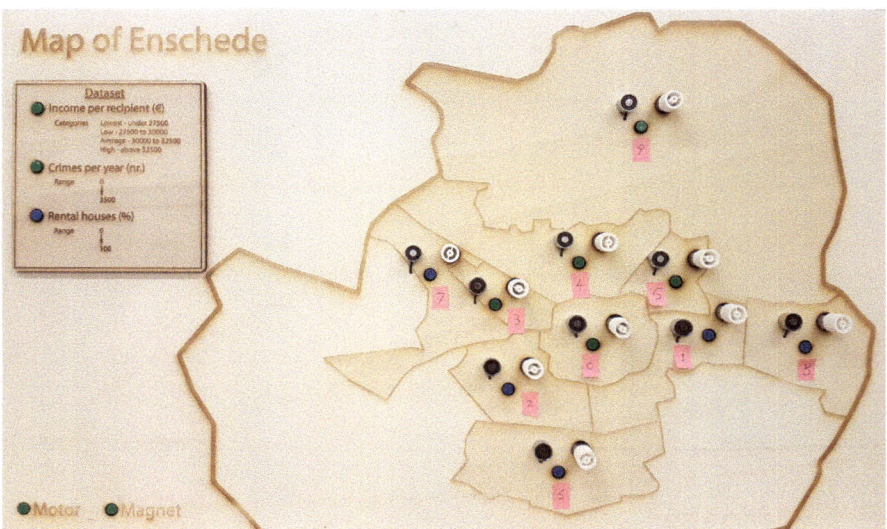

Figure 4-2. Implementation of a thematic map using the haptic kinesthetic variables of resistance and friction. The magnet is at the left and the spinning wheel is at the right of each neighborhood's symbol group. Image courtesy of Sander Dullaert.

two variables (resistance or friction) to explore how one of three numeric variables (incomes, crime rates, or percentage of houses that are rentals) vary across a city's neighborhoods. Resistance is produced by a user bringing a metal handle near a magnet located within each neighborhood.[3] A larger data value produces a stronger pull toward the magnet. Friction is generated through a spinning wheel that the user touches and that has a rough surface. A higher spinning speed generates more friction.

Not all haptic variables have a directly analogous visual variable, and not all visual variables have a directly analogous haptic variable. Most conspicuously, the visual variables derived from components of color (hue, lightness, and saturation), visual variables that are used frequently on visual maps, are missing. Highly contrasting colors can be included on tactile maps because they are useful for those tactile map readers with residual vision.[4] To create legible maps for those map readers with no vision, it may be possible to transform color to a tactile variable. For example, some researchers have suggested that colors could be transformed into geometric shapes (figure 4-3) or textual codes communicated in either braille or

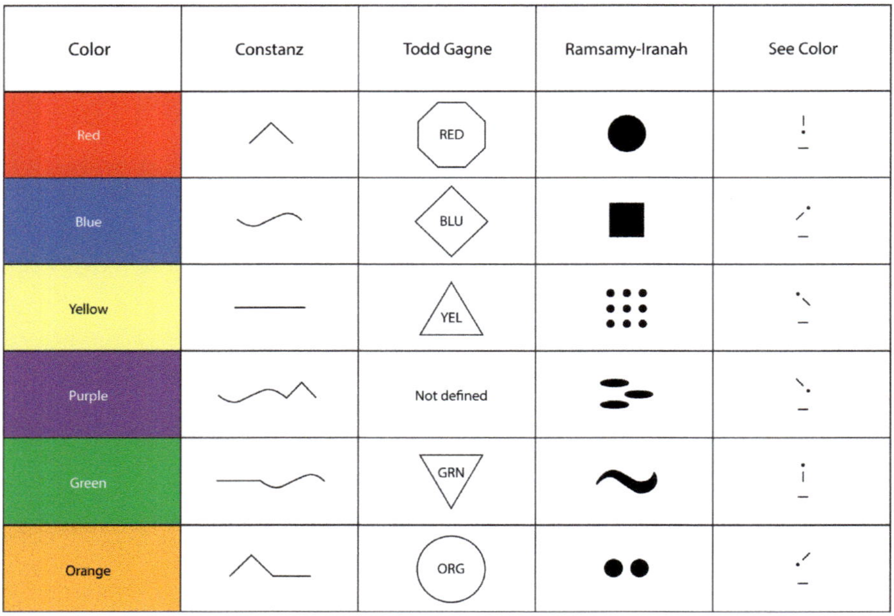

Color	Constanz	Todd Gagne	Ramsamy-Iranah	See Color
Red	∧	RED	●	!
Blue	∼	BLU	■	′
Yellow	—	YEL	⁝⁝	‵
Purple	⌢⌒	Not defined	⬬⬬	‵·
Green	⌒	GRN	∼	i
Orange	∧	ORG	••	′

Figure 4-3. Tactile symbols used to communicate color hues in the Constanz, Todd Gagne, Ramsamy-Iranah, and See Color symbol sets. Adapted from de Araújo et al. (2020), CC-BY.

Latin characters, all of which are perceivable by map users with no vision. Preliminary experiments show that this approach for transforming color can be more successful than the use of different fill textures and may be a good alternative to color hue,[5] although the grammatical logic of color lightness as ordered may not translate so effectively, particularly for those map readers whose vision impairment is congenital.[6]

Explanation, exploration, and comparison of visual and haptic variables

Chapter 3 ("Understanding through touch") highlights some of the differences between how people perceive the world through the senses of vision and touch. Here, some of these differences are discussed further, with reference to the visual and haptic variables that mapmakers can use to create symbols.

Vision is a high-acuity sense. We see with great detail in our focal vision, in the locations where our eyes are pointed. This means that we can detect small differences in shape and color, and this capability allows mapmakers to develop symbols that have relatively small differences yet that can be distinguished by the map user. This is possible because we have a very high density of cells in our eyes specialized in sensing light. Touch, on the other hand, is a lower-acuity sense compared with vision. This means that the difference between two tactile symbols must be greater than between two visual symbols for the map reader to perceive two symbols as different from each other. It also means that the density of information that can be communicated with touch is lower than with vision. Chapter 6 ("Generalization for Tactile Maps") explores how the acuity of each sense (vision and touch) drives some of the constraints required for map generalization, to ensure that tactile maps are legible. For example, according to the *Guidelines and Standards for Tactile Graphics*, published by the Braille Authority of North America in 2022,[7] the minimum distance between two haptic symbols that is required for each to be perceivable as an individual symbol is one-eighth of an inch (a little more than 3 mm), compared with a 0.1 mm separation that is required for vision.

Tactile sensations are produced by mechanoreceptors, which are sensitive to pressure, or thermoreceptors, which are sensitive to temperature. These receptors exist at varying densities throughout the skin, with the highest densities in the hands, meaning that the highest tactile acuity for complex shapes comes from using your hands to explore a surface. Both the human eye and the human hand require movement to put together a picture of the world. The sensory information processing parts of the brain stitch together sensory information from individual snapshots to produce this integrated picture.

A difference, however, between vision and touch is the field of view of each snapshot. Binocular human vision (that is, using both eyes together) has a field of view of around 214 degrees without any eye movements. Touch, meanwhile, has a field of view for each snapshot that is limited by the size of (at most) two hands put side by side, an angle that will vary between people but is in any event much smaller than 214 degrees. This means that the brain must stitch together many more snapshots from the hands than the eyes to produce a complete picture of a map, and this requires more reliance on working memory, which is a limited resource in the brain. This also means that it may be harder to recognize the meaning of iconic symbols through touch than with vision, so tactile map readers must rely on the legend. Moreover, as discussed in chapter 3, haptic perception can be influenced by the order in which elements are explored because of perceptual aftereffects.

Together, these limitations constrain how the mapmaker can make map symbols that function well using visual or haptic variables. Yet, the degree to which they can do so might also depend on other aspects of map design, such as the scale and size of the area being mapped as well as the physical size of the map itself. With visual variables, at least for most static maps, it is possible to see the entire map at once to get an overview of the location or pattern, even if inspection of the map's details might require the reader to focus their attention on different areas of the map. Interactive maps that allow the reader to pan and zoom show more limited sections of the map at one time, which also means that they must stitch together multiple snapshots to understand a location. Map readers can sometimes get lost within these interactive maps. One proposed solution to this problem, anchors,[8] is, in fact, similar to an anchoring strategy that is sometimes used by people who are reading tactile maps and graphics. Anchors are fixed objects (an always-drawn landmark in visual maps or one hand that doesn't move in tactile maps) that reduce disorientation by providing a known reference point. This implies that future cross-modality (vision and touch) research might provide solutions that benefit all map readers, whether they consult visual or tactile maps.

Like visual maps, maps that use haptic symbols can be produced using different technologies, including both physical and digital formats. The production technology can affect a symbol's rendering quality, meaning that this should be a factor that informs the mapmaker's choices. An example of this in visual maps is that an icon that is optimized for screen display may appear pixelated or not detailed enough when printed on paper. Haptic symbols are similarly affected by the capabilities of the chosen production technology. Although the legibility of haptic symbols has not been studied systematically across every production technology, some research that examined a set of previously validated tactile symbols has concluded that well-designed tactile symbols can be functionally equivalent across multiple production technologies. This research also found that some production technologies (3D printed symbols) created maps that could be read more quickly by their users than the other production technologies examined (microcapsule paper and embossed symbols).[9] An important recent trend in tactile mapping is the proliferation of new production approaches, which make it easier to generate interactive, multimodal tactile and haptic maps. Multimodal maps, for example, can help solve one of the most important problems of tactile map design (needing more space for symbols) by enabling map labels or the meaning of legend symbols to be communicated on demand through speech rather than being printed in raised Latin letters or braille.[10]

Capabilities and use of variables for map design for different purposes

Mapmakers who make visual maps often make them to communicate an important message, such as which locations are at risk from a hazard, such as flooding. But at other times, they might make maps to explore a dataset and to discover something new about a geographic phenomenon. There is no reason why haptic maps cannot support both tasks well, if symbols are chosen appropriately for the map's purpose.

Maps for communication versus maps for exploration

A primary requirement for maps that are designed for communication is that the map symbols for different themes represented on the map are unambiguous and clearly differentiable. In visual maps, although redundant use of two visual variables (using both variables to represent the same data theme) is a design choice sometimes made by the mapmaker, more often two different visual variables are used to represent two different data themes. In tactile maps, multiple haptic variables might need to be employed to make a given set of symbols differentiable, meaning that two haptic variables are needed for one data theme. For example, recent research evaluating the recognition of a set of point symbols found that varying only one haptic variable

Figure 4-4. Tactile symbols that differ by one haptic variable, tangible orientation (*left*), or two haptic variables, tangible orientation and tangible shape (*right*). Redundant use of two haptic variables increases the tactile discriminability of the two symbols.

such as symbol orientation (rotation) was not enough to produce reliable recognition (figure 4-4).[11]

Although many maps are used for communication purposes, maps are also used for exploration and discovering new patterns and geographic relationships. People with vision impairments have been largely excluded from using maps constructed for these purposes. Many fewer thematic tactile maps have been produced compared with the larger number of maps that support understanding and navigating through local environments. Nevertheless, some features of haptic variables can support exploring data themes. Some haptic variables are particularly salient (see chapter 3), having perceptual effects similar to the visual "pop-out" effect that occurs with color, in which an object of contrasting color is immediately noticeable to the map reader. Some examples of haptic variables that can be used this way are texture (roughness), resistance (as in the map depicted in figure 4-2), and temperature. These might be haptic variables that assist map users with data exploration tasks, such as change detection or detecting outliers.

Another common data exploration task involves comparing the spatial distributions of two data themes. Recent preliminary research has shown that when chosen well, haptic variables can support combining two data themes on a single map (tangible elevation and tangible numerosity), with similar time needed to explore the map and similar accuracy in understanding the map patterns on the multivariate map compared with two individual maps of the data themes.[12]

Navigation maps versus thematic maps

Large-scale maps that are used for navigation and wayfinding help their users build a rich mental map of the relationship of different entities contained within the map's extent, such as streets, buildings, footpaths, intersections, and even potential routes from one location to another. Further, they can provide context for vocal descriptions of routes that might come from a navigation app and help map users know what to expect while navigating a route. Thus, they can be of particular assistance for route planning.

Tactile navigation map design is challenged by space because map labels, which are necessary for understanding the names of features such as streets, buildings, neighborhoods, and so on, must be included on navigation maps for them to be useful. A physical map can be made larger to have more space for showing map features clearly, but this is at the expense of portability. Moreover, tactile maps printed on microcapsule paper or thin plastic maps can be difficult to read while mobile because they need the support of a flat surface and continuous exploration by unoccupied hands. Other technologies for controlling the exploration of mapped space are available, such as videogame controllers. This option frees the designer from the limitations of the small size of a touchscreen and adds the possibility of producing haptic variables, such as vibration amplitude or vibration frequency. Another benefit of these technologies is that they can also be paired with other sensory modalities, such as hearing, to create an audio-haptic map. Such maps can divide information between sensory modalities, such as using audio for communicating place-names, so that other information can be communicated with haptic variables.

A key challenge in creating tactile thematic maps is that it can be harder to establish the intellectual hierarchy on the map because there is often a need to generalize and reduce the total amount of information contained on the map so that it is legible. In a visual map, less important features, such as basemap features that provide context, would be visually deprioritized with visual variables, such as lightness and size. For example, less important labels would use a color lightness that had less contrast with the map background than more important labels, and less important streets might use finer lines. One design alternative for a tactile thematic map is to construct two maps to be read side by side, one showing only the basemap and the second containing the thematic content plus the basemap content.[13] Another possibility, when the production technology allows (for instance, 3D printing), is to use the haptic variable tangible elevation to literally elevate the most important information.[14]

Conclusion

This chapter introduced the concept of cartographic symbol grammars and described similarities and differences between what mapmakers can manipulate to construct map symbols when creating visual and tactile maps. To create legible symbols, mapmakers need to be informed about the different levels of acuity of vision and touch and other relevant contextual factors, such as the size and scale of the map, the map production technology, and the map's purpose.

Further reading

On cartographic symbol grammars

Bertin, Jacques. *Semiology of Graphics: Diagrams, Networks, Maps*, 1st ed. Esri Press, 2010.

de Araújo, Niédja S., Luciene S. Delazari, V. O. Fernandes, and Mauro A. Júnior. "A Bibliometric Study of Graphic Variables Used on Tactile Maps." *International Archives of Photogrammetry and Remote Sensing and Spatial Information Science* XLIII-B4-2020 (2020): 25–32.

Dullaert, Sander, Auriol Degbelo, Champika Ranasinghe, and Nacir Bouali. "Data Physicalization with Haptic Variables: Exploring Resistance and Friction." In *CHI EA '24*, May 11-16, Honolulu, HI (2024): 1–8.

Griffin, Amy L. "Feeling It Out: The Use of Haptic Visualization for Exploratory Geographic Analysis." *Cartographic Perspectives* 39 (2001): 12–29.

MacEachren, Alan M. *How Maps Work: Representation, Visualization, and Design*. Guilford, 1995.

Wabiński, Jakub, and Emilia Śmiechowska-Petrovskij. "Evaluation of Qualitative Colour Palettes for Tactile Maps." *ISPRS International Journal of Geoinformation* 13 (2024): 94.

White, Travis. "[CV-03-008] Symbolization and the Visual Variables." In *The Geographic Information Science & Technology Body of Knowledge* (2nd quarter, 2017 edition), edited by John P. Wilson.

On explanation, exploration, and comparison of visual and haptic variables

Ranasinghe, Champika, and Auriol Degbelo. "Encoding Variables, Evaluation Criteria, and Evaluation Methods for Data Physicalisations: A Review." *Multimodal Technologies and Interaction* 7 (2023): 73.

Yoshida, Takako, Ayumi Yamaguchi, Hideomi Tsutsui, and Tenji Wake. "Tactile Search for Change Has Less Memory than Visual Search for Change." *Attention, Perception, and Psychophysics* 77 (2015): 1200–1211.

On capabilities and use of haptic variables for map design for different purposes

Cole, Harrison, and Anthony Robinson. "Thematic Tactile Maps for Accessible Flood Mitigation Planning: Design and Evaluation." *Cartography and Geographic Information Science* 50, no. 6 (2023): 574–92.

Wabiński, Jakub, Albina Mościcka, and Marta Kuźma. "The Information Value of Tactile Maps: A Comparison of Maps Printed with the Use of Different Techniques." *The Cartographic Journal* 58, no. 2 (2021): 123–34.

About the author

Amy L. Griffin is a senior lecturer, School of Science, RMIT University, Melbourne, Australia, and vice president of the International Cartographic Association. Amy completed a BSc in geography from Macalester College in the United States and an MSc and PhD in geography from Penn State University in the United States. After her PhD, she moved to Australia, first working as a lecturer at the University of New South Wales in Canberra, then moving to Melbourne in 2017. Her research interests range across cartography, but most of her projects include a focus on the end users of maps and how they read and understand maps. Her interest in mapping with nonvisual methods began during her MSc studies when she wrote a paper on haptic variables. In her spare time, she plays the banjo and mandolin.

Amy L. Griffin.

Notes

1. Bertin, Jacques, *Semiology of Graphics: Diagrams, Networks, Maps* (Esri Press, 2010).
2. Griffin, Amy L., "Feeling It Out: The Use of Haptic Visualization for Exploratory Geographic Analysis," *Cartographic Perspectives* 39 (2001): 12–29.
3. Dullaert, Sander, Auriol Degbelo, Champika Ranasinghe, and Nacir Bouali, "Data Physicalization with Haptic Variables: Exploring Resistance and Friction," in *CHI EA '24*, May 11-16, Honolulu, HI (2024): 1–8.

4. Wabiński, Jakub, and Emilia Śmiechowska-Petrovskij, "Evaluation of Qualitative Colour Palettes for Tactile Maps," *ISPRS International Journal of Geoinformation* 13 (2024): 94.
5. Loch, Ruth Emilia Nogueira, "Cartografia Tátil: Mapas para Deficientes Visuais," *Portal da Cartografia* 1, no. 1 (2008): 35–38.
6. de Araújo, Niédja Sodré, et al, "Avaliação do Sistema de Código de Cores 'See Color' em Mapa Tátil," *Revista Brasileira de Cartografia* 72, no 1 (2020): 34–48.
7. Braille Authority of North America, *Guidelines and Standards for Tactile Graphics, 2022* (Braille Authority of North America, 2022).
8. Touya, Guillaume, María-Jesús Lobo, William A. Mackaness, and Ian Muehlenhaus, "Please, Help Me! I Am Lost in Zoom," *Proceedings of the International Cartographic Association* 4 (2021): 107.
9. Brittell, Megen E., Amy K. Lobben, and Megan M. Lawrence, "Usability Evaluation of Tactile Map Symbols Across Three Production Technologies," *Journal of Visual Impairment & Blindness* 112, no. 6 (2018): 745–57.
10. Cole, Harrison, "Tactile Cartography in the Digital Age: A Review and Research Agenda," *Progress in Human Geography* 45, no. 4 (2021): 834–54.
11. Mościcka, Albina, Emilia Śmiechowska-Petrovskij, Jakub Wabiński, Andrezj Araszkiewicz, and Damian Kiliszek, "Methodical Testing of Tactile Cartographic Signs in Isolation and in Context," *Cartography and Geographic Information Science* (2024): 1–18.
12. Zacharogiorga-Sourdi, Alkistis, Margarita Kokla, and Eleni Tomai, "Evaluating the Usability of 3D Thematic Maps: A Survey with Visually Impaired Students," *British Journal of Visual Impairment* 41, no. 3 (2023): 646–61.
13. Cole, Harrison, and Anthony Robinson, "Thematic Tactile Maps for Accessible Flood Mitigation Planning: Design and Evaluation," *Cartography and Geographic Information Science* 50, no. 6 (2023): 574–92.
14. Wabiński, Jakub, Albina Mościcka, and Marta Kuźma, "The Information Value of Tactile Maps: A Comparison of Maps Printed with the Use of Different Techniques," *The Cartographic Journal* 58, no. 2 (2021): 123–34.

Chapter 5

Map design and cognition

Jakub Wabiński and Simon Ungar

Traveling has always been my greatest passion. It all began when I was just five years old, and my parents took my brother and me on our first holiday in the mountains. For a hiking trip, maps are essential, and I still remember begging my parents for my own copy of the map, eager to navigate the trails by myself. Of course, as a result, I sometimes got lost. This early fascination with maps sparked a lifelong love that continues to this day. Years later, while I was working at a small photogrammetry company, my boss asked me to repair a 3D printer that a client had returned. I had no clue where to start, but that challenge ignited my interest in additive manufacturing techniques. When I returned to my studies, I realized that 3D printing could be perfect for creating raised 3D maps. But wait, aren't those the maps used by people with visual impairments? It was only later that I discovered they are called tactile maps… and I'm still fascinated by them.

—Jakub

I have always been fascinated by the beauty and power of maps and can remember being engrossed by maps of fantastic lands in *The Chronicles of Narnia* books and *The Hobbit*. When traveling with my family, I always asked to be the map reader, whether traveling across Europe in the car or walking in the Pentland Hills near our home in Edinburgh, Scotland. While studying for my first degree in psychology, I became intrigued by lectures on spatial cognition. Then, when I was offered the chance to stay on and do a PhD on the potential of tactile maps to develop the spatial understanding of blind children, I quickly accepted, maintaining my interest and focus in this area for several years. Since switching careers to become an educational psychologist, I have continued my professional involvement with blind and vision-impaired children.

—Simon

Map creation stages

Before we begin our consideration of tactile map design, we should note that map design can refer both to the process of creating maps and the effect of their creation—the map itself and the way it is designed. Cartographic knowledge is crucial for designing good maps, but even the most skilled cartographer might fail to produce a legible and accurate map without consulting domain experts and target user groups. This is particularly important for thematic maps and essential for tactile maps.

Regardless of the map type, the map creation process can be divided into three main stages (figure 5-1): conceptual, preparatory, and production.

First, at the conceptual stage, a problem has to be identified—that is, the *what*, *where*, and *who* of a map: in other words, the topic, geographic reference, and target user group. These determine map content, scale, format, cartographic visualization methods, variables used, production techniques, and publishing approaches.

Next, data must be collected and evaluated. Do I need to gather data or reuse the existing data? Do I have to georeference paper maps (using coordinates to tie maps or images to real-world features)? Are they of sufficient quality? After that, data must be processed, which involves cleaning, transforming, and generalizing it—this last being crucial for tactile maps, considering the limitations of tactile perception compared with vision. Later, map layout is defined, symbols are designed, and their hierarchy set.

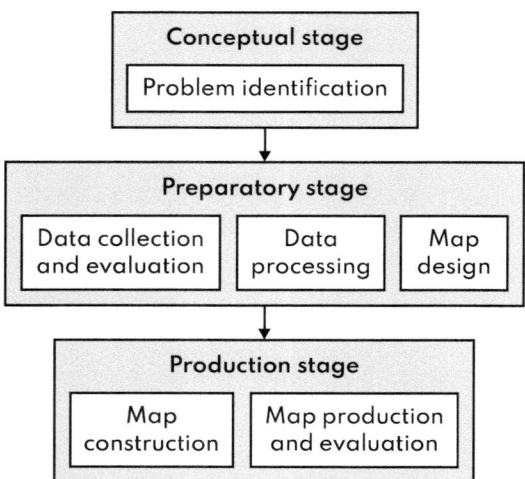

Figure 5-1. Map creation stages.

Finally, in the production stage, the map is constructed—that is, created using cartographic tools and techniques based on previous assumptions. In the last step, the map is printed (in the case of analog products) or published (digital products). This process should involve test prints and proofreading (user testing).

In this categorization, map design is only a small part of the overall map creation process, focusing on symbol selection and layout. However, in this chapter, we will take a holistic approach, considering how map design affects perception and how user perception influences map design, bearing in mind the cognitive capabilities of people with visual impairments.

Map design

Map design is a complex process, and each map is unique, so there is no universal recipe for a good map. Still, cartography has long traditions, and mapmakers can follow best practices developed over the years. Some rules derive from established theories, such as Peirce's theory of signs (semiotics),[1] Gestalt design principles,[2] and principles derived from cognitive psychology.[3] Based on visual cartographic approaches, five equally important factors determine the map design process: the nature of geographic phenomena, the recipient, technical limitations, map scale, and the purpose of the map.

In tactile cartography, some factors are more important than others. The *nature of geographic phenomena*, interpreted through spatial relationships, is heavily influenced by the perceptual and cognitive capabilities of the map user (*recipient*). For people with visual impairments these capabilities are different from those of sighted users and in some respects more limited. Additionally, tactile cartography involves physical relief maps, which are more complex to produce than digital maps because of *technical limitations*. These maps cannot be too small, because tactile signs and braille characters require more space than their visual counterparts, but they also cannot be too large, because all elements must be within the reader's arm's reach. All this influences the *map scale*.

But it is the *purpose of the map* that influences all the other factors. What is the map's topic? What area will it cover? Who will use it? How and where? All these questions should be answered before the map production stage, bearing in mind the map reading process.

The map reading process has three levels. First, noticing and understanding signs. Second, interpreting relationships among map features. Third, analyzing and synthesizing spatial information to draw conclusions. Cartographic visualizations

Table 5-1. Rules to enhance map perception

Rule	Description
Sufficient legibility	Adequate size and spacing of signs
Distinguishability	Sufficient contrast between signs obtained using graphic and haptic variables
Ease of recognition	Legible and consistent symbology
Balance and aesthetics	Uniform style, harmonious colors and textures, balanced composition

are powerful tools but they rely on a reader being able to distinguish and recognize map elements before being able to draw conclusions. Tactile map designers must consider readers' capabilities, especially the perceptual limitations of touch and residual vision. Although visual and tactile map design principles differ, some universal rules enhancing map perception can be identified (table 5-1).

Yet again, these rules must be reconsidered for use in tactile cartography because of the nature of tactile perception. The cognitive turn in cartography has provided a helpful set of theoretical and practical frameworks for integrating many of the issues we've discussed so far. In the next section, we provide an overview of this field and of its subsequent extension into tactile cartography.

Map cognition

Historically, the design of maps has been practice-led, with a focus on the primary task of accurately rendering spatial data visually in a way that is aesthetically pleasing. Decisions about map symbols and their deployment across the map's surface were made by the artists or specific professionals (engineers, soldiers, civil servants) tasked with creating maps for different purposes. Although attention was paid to aesthetics and legibility in a broad sense, it was not until the second half of the 20th century that cartographers began to seek a more scientific approach to map design, drawing on psychological theory and research to inform map design. This, in turn, brought a user-centered, "human factors" approach to the field.

The cognitive turn in visual cartography

Following cartographer Arthur Robinson's call in 1952 to apply human perceptual and cognitive factors to the design of maps (see Montello, 2002), several researchers embarked on attempts to bring knowledge, theory, and research methods from psychology into the map design process. The focus was initially on the evaluation of map symbols to perform specific functions (for example, the ability of thematic map symbols to support quick and easy comparison between mapped properties). Later, designers drew more directly on research in psychophysics and cognitive psychology to understand how cognitive and perceptual processes of visual attention and memory affect the detection, discrimination, identification, and interpretation of cartographic symbols and their spatial arrangements.

The aim was not to erase centuries of accumulated practice-based knowledge and understanding of the art and craft of cartographic design, but rather to apply psychological theory and research methods to improve understanding of the way maps work, to enhance the quality of map design, and to understand why some maps may be more effective than others.

Complex models of cartographic communication were developed, linking mapped reality through the cartographic design process via the map to the map user and their resulting understanding or image of the world (figure 5-2). Among other things, these models emphasized the significant role of perception (of the world and the map) and cognition (real-time processing of the map and integration of map information with other information stored in memory). They sought a grammar of map design based on empirical research into the perception and cognition of map symbols and their combinations.

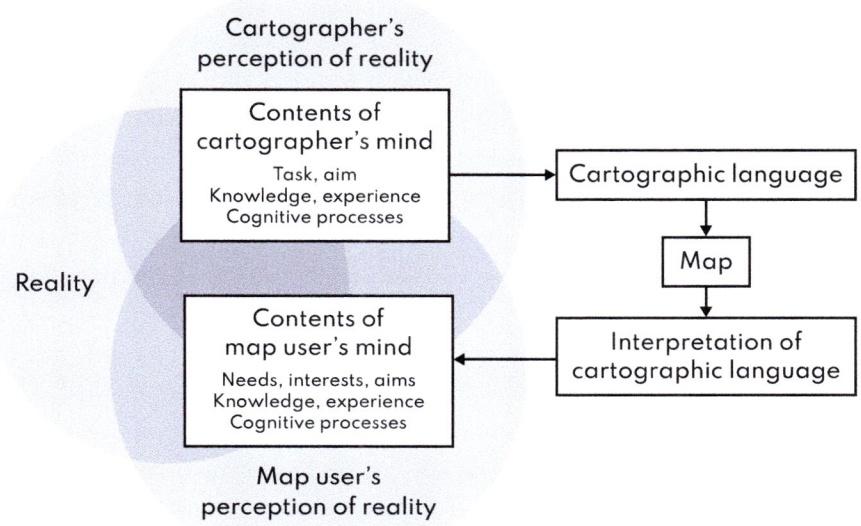

Figure 5-2. Cartographic communication model—map user's perception of reality might differ from that of a cartographer, potentially leading to misunderstandings. Adapted from Koláčný.[4]

Cognition in tactile cartography

If this approach to the design of visual maps can be called cognitive visualization, it has been suggested that a parallel "cognitive tactualization" approach might be applied productively to the design of tactile maps. Although existing practice-led tactile map design guidelines are likely to contain important insights into key aspects of map legibility, they may also include aspects that are not optimal for users. The cognitive tactualization approach aims to take practice-led design guidelines and existing tactile symbol sets as a starting point from which to carry out evaluation of their effectiveness in structured testing with participants with visual impairments.

When testing the legibility of tactile map symbols, several fundamental perceptual and cognitive principles are relevant. At the most basic level, individual symbols must be detectable (see saliency of features in chapter 3, "Understanding Through Touch"). This means that they must stand out from the background on which they are based (for example, insufficient elevation above the base might mean that the symbol is not detected at all), discriminable (that is, reliably differentiated from other tactually similar symbols), and identifiable (that is, the symbol is reliably identified and associated with its intended referent or meaning). The capacity of the fingertip to accurately detect, identify, and discriminate different forms and textures is significantly lower than for vision, meaning that far fewer forms and patterns are available to the tactile map designer. This also means that there is less scope for "redundancy" (that is, the combination of two factors, such as form and texture, to uniquely identify a symbol) in the design of tactile symbols. Many studies have been carried out to explore and define sets of maximally discriminable point, line, and area symbols. As a rule, the "palette" of reliably and easily discriminable signs for tactile symbol design is much more limited than for visual symbols. This means that a limited pool of symbols will need to be reused to denote different things on different maps.

A question at a higher cognitive level is the extent to which a map symbol is designed to resemble its referent or some aspect of it, as opposed to a more abstract symbol that can be linked to the referent only through the map key. Although there are advantages in resemblance, in practice this can be difficult to achieve in a reliable way, even on visual maps. A symbol that resembles its referent visually may not serve the same purpose in a tactile form—for example, a triangle used to represent a mountain or wavy lines in an area symbol used to denote water both bear a visual resemblance but are unlikely to correspond to the way these environmental features are experienced by people with visual impairments. They will therefore be, for all intents and purposes, abstract symbols. Because of the way people with visual impairments perceive the world through the nonvisual senses, it is difficult to identify tactile representations that resemble large-scale environmental features in any immediately meaningful way. Also, because of the more limited resolution (acuity) of touch relative to vision, the complexity required to render most pictorial symbols will make it impossible for the resemblance to be reliably identified or recognized by touch—for example, a symbol resembling a house is likely to be perceived as a textured or rough square. There are a few physical resemblances that can be exploited, such as a rough sandpaperlike texture to represent a sandy beach, and there are some conventional symbols that would work well in the tactile format—for example, a letter *P* for parking. However, in general, tactile symbols need to represent their referents in a more abstract yet conventional way, relying on a key or preteaching to allow the reader to draw meaning from a map.

The capacity to perceive tactile detail also depends on the orientation of the symbol in interaction with the scanning action of the reader's fingertip. For example, it is easier to distinguish single and double lines when they are oriented vertically as opposed to horizontally.[5] This is most likely due to the tendency for readers to move their fingertips from side to side when scanning the lines, providing greater perception of shear when lines are perpendicular to the direction of scanning. If readers were trained to scan in different directions, this effect might vanish.

Data produced in psychophysical studies of tactile symbols in isolation may act as a helpful guide; however, they are not a sufficient basis for guidelines for the design of complex tactile maps where a variety of symbols vie for the reader's attention and where it may not be immediately obvious if a dot is a point symbol, part of a line or area symbol, or part of a braille letter. Although attempts have been made to test the findings from psychophysical studies in the context of whole maps (see, for example, notes 3 and 6), this important area has been underresearched. As has been noted in research on visual maps, the number of variables that need to be considered and potentially controlled in such studies makes them extremely difficult to design.

Global and local perceptual processing

When sighted map readers see a map, they almost immediately register the shapes of different areas or regions of the map as well as global patterns and configurations of symbols. This is due to rapid simultaneous visual processing that registers a range of features and structures in a scene before any more conscious and deliberate sequential visual searching has taken place.[6] For example, bright-red point symbols on a map are likely to pop out of the map surface, and where there are several of these (for example, representing capital cities across a regional map), the spatial configuration of these points will be perceived (according to the Gestalt principle of similarity[7]). This "preattentive" processing of a visual map is automatic (that is, not consciously controlled by the reader) and serves to prime the reader to expect certain kinds of information in certain parts of the map. It thus acts as a guide to subsequent, more deliberate, and conscious visual search of the map.

Research on this more conscious and deliberate visual attention and visual search has likened the process of engaging with and extracting information from a visual scene to the zoom lens of a camera. The focus of our attention can be shifted from an entire scene to zoom in on successive regions of interest across the scene. In tactile map reading, the perceptual engagement with the map is very different, consisting of a sequence of encounters with different symbols and parts of (line and area) symbols, in a sense analogous to a permanently "zoomed in" form of visual attention or like examining a scene through a tube. The fingertips of both hands are potentially available to do this, although in practice the nondominant hand is often used as an anchor or reference point to keep the moving hand oriented relative to the frame of the map (for example, by resting at one corner of the map). Experienced map users may develop the skill of using the fingertips of both hands as separate "lenses" on the map surface.

It is therefore only after a great deal of exploration of the map that tactile readers build up impressions of the shapes of regions, networks of lines, and the relative positions of point symbols. Where the map has a separate key, readers will also have to divide their attention between this and the map surface as they locate and identify individual symbols. This also inevitably places increased demands on working memory, as a tactile map reader gradually pieces together and integrates the successively experienced impressions of the map. The approach they take and strategies they bring to this task will determine the effectiveness with which they extract and synthesize meaningful information from the map. Therefore, a compromise will always need to be made between the information value of a map and the processing capacity of the map user, bearing in mind that more experienced map users will make more efficient use of limited perceptual and cognitive processing resources. Researchers have considered ways in which information about global grouping or structure can be provided in tactile form—for example, by using a separate outline map to provide information about global structure before or alongside a more detailed map.[8] However, this approach has been relatively underresearched.

Experience and training

A map can be designed in many ways. Because of different perceptual capabilities among people with visual impairments, designing an informative tactile map that is universally legible and comfortable is a challenging task. People with visual impairments who lost their sight at or soon after birth are likely to form mental representations of the world around them that are different in structure and content from those of sighted people or those who lost their sight later in life. Many people with visual impairments have some level of vision, so hybrid maps with tactile and simplified graphic content should be designed whenever possible.

People with visual impairments generally have much less access and exposure to tactile maps and graphics than sighted people have to visual materials, so you cannot make the same assumptions about levels of experience and understanding of tactile graphics as you might about a sighted person encountering visual material. There are large variations between people with visual impairments in the level of map-reading skills, braille literacy, and openness to technology, which emphasizes the need for training in basic aspects of what Frances Aldrich has termed "tactile graphicacy."[9] When people are asked to extract spatial information from a tactile map, they differ in their approach to the task (that is, they use a variety of map reading strategies to read a map and to synthesize the information in the map), in ways that determine how effectively they make sense of and use the information from the map. Therefore, the best-designed map will only be as effective as the map-reading experience and skill of the reader allow.

With all this in mind, we want to emphasize the importance of a participatory approach in tactile graphics design. Users' needs may differ from designers' expectations (see figure 5-2). For instance, sensory clues such as ticking crosswalk signals are crucial for people with visual impairments on orientation and mobility maps but are often overlooked. Furthermore, user evaluation should not be restricted to the basic effectiveness of a map but should include aspects of comfort and aesthetic qualities. People with visual impairments have just as much right as sighted people to maps that are attractive and comfortable to use (see chapter 7, "User-Centered and Inclusive Cartographic Design"). Ideally, solutions should be iteratively tested by sociodemographically diverse groups of people with visual impairments to ensure their usefulness and comfort. Well-designed tactile maps must balance user requirements, which may be mutually contradictory, with practical production aspects, including cost, time, and complexity.

Best practices in tactile map design

Legibility is of the highest priority on tactile maps. Thus, following tactile graphic design principles is more important than adherence to strict mathematical and cartographic rules. For example, maintaining minimum distances and preserving topological relations is more important than precise location accuracy of the mapped features. Features can be generalized as long as the original intent, determined by the map designer, remains intact. This section contains several recommendations and examples of good practice derived from relevant literature and less official local guidelines. They have been empirically verified by one of the authors in user studies with people with visual impairments, using methods based on the cognitive tactualization approach. For more detailed and parametrized guidelines, see the "Further Reading" section. All the guidelines presented here are meant for general-purpose maps and may not be applicable for more specific, custom-made maps.

Symbology

When designing symbols on tactile maps, it is recommended wherever possible to use simple geometries that resemble the mapped features in ways that are meaningful for the user with visual impairments. This will aid understanding and memorization. Raised symbols are easier to trace and recognize than concave ones. Symbols must be easily distinguishable, the way that alphabet letters are. This can be achieved, whenever possible, through redundancy—that is, by varying symbols by more than one graphic or haptic variable (figure 5-3). Height differentiation, although only available with selected production techniques, facilitates symbol identification on tactile maps. As a rule of thumb, smaller symbols should be raised higher to facilitate their identification, with braille letters raised to a unique height, which is determined by the existing standards.

Maps use three main geometries: points, lines, and areas, with labels being a specific additional symbol type. Point symbols depict real-world features that are

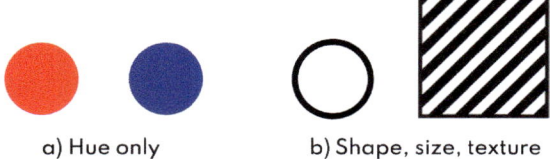

Figure 5-3. The more variables involved to differentiate symbols, the better. In (a) only hue was used, and in (b) three variables are used, which greatly increases their distinguishability.

too small to be presented with lines or areas. They should fit under the reader's fingertips while being large enough to ensure legibility. Sharp, clean-cut edges attract more attention than rounded ones. Larger point symbols are easier to discriminate as outlines, while smaller ones should be solid.

Lines represent features whose length can be expressed in the map scale. For tactile maps, four aspects of line design can be identified: single-double, continuous-broken, smooth-ragged, and thick-thin. Single and broken lines are easier to trace, but broken lines should allow the reader to feel the next segment before leaving the first. Thick and ragged lines are easier to identify but might be harder to trace.

For large features, where both location and extent matter, area symbols (textures) should be used. Textures can clutter a map visually or haptically, making it hard to identify other symbols. Rough textures attract more attention. Tracing large objects requires building a mental image, making it difficult to conceptualize enclosed areas. Thus, it is important to limit not only the number of unique area symbols but also the total area they cover on maps.

Bear in mind that the capacity of the fingertip to accurately detect, identify, and discriminate different forms and textures (haptic resolution) is significantly lower than for vision, meaning that far fewer forms and patterns are available to the tactile map designer. Conditions such as diabetes or prolonged disinfectant use can further reduce touch sensitivity.

Labels convey information that cannot be presented graphically. Tactile maps use braille together with large-print text, both requiring a considerable amount of space. Labels should be placed horizontally and in straight lines (unless they need to be placed along linear features). Labels should not obscure important map details but at the same time must be repeated along longer line features so that they are identified correctly. Abbreviations and codes are common, although abbreviations may be easier to associate with mapped feature names. Both should be explained in a legend or separate sheet.

Map miscellanea

A tactile map must include not only the map area itself but also the title, legend, orientation mark, and imprint describing data timeliness and authorship. Inset (locator) maps should be avoided because they are hard for people with visual impairments to understand. The inclusion of a scale and north arrow is debatable on tactile maps because they might be too abstract for people with visual impairments.

The map legend must be complete, explaining every symbol used (figure 5-4). Symbols should be ordered hierarchically with the most important thematic features on top or grouped by geometry with point symbols first, followed by lines and areas. Letters and numbers should be in alphabetical and numerical order. On large-scale maps, orienting symbols, such as "you are here," should appear first. Despite limited braille literacy among people with visual impairments nowadays, legends should nevertheless include both braille and large-print descriptions. Point symbols in a legend must match those on the map. For line symbols, it is recommended to use straight line segments for human-made features and curves for natural ones, while rectangles should be used for representing area symbols. All symbols must be large enough to be legible by people with visual impairments. Minimum recommended dimensions of signs differ depending on the context. A map legend can either be part of the map sheet or separate. The first approach requires no additional

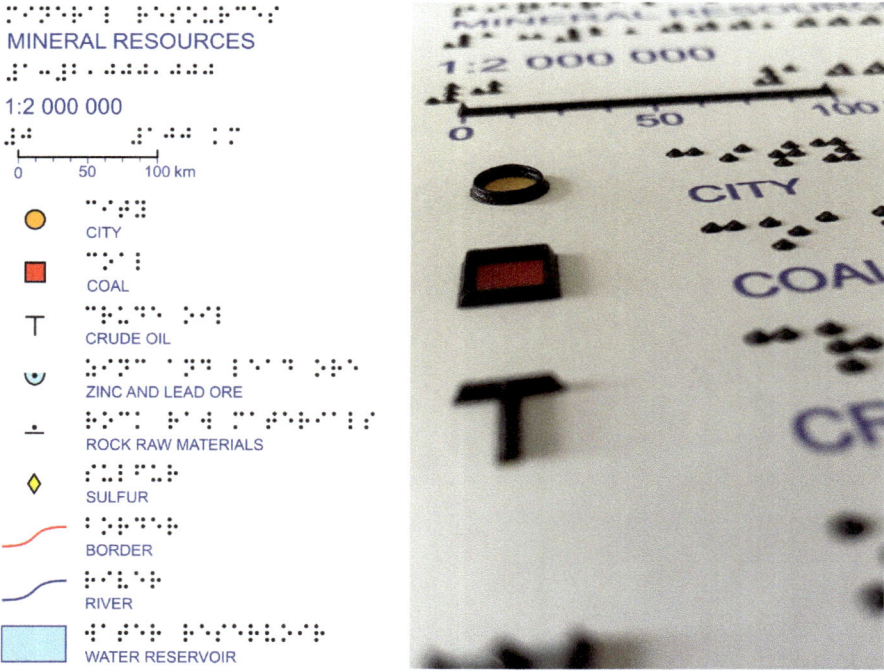

Figure 5-4. Thematic map legend example—digital model on the left and UV printed output on the right. Symbols aligned to the center with descriptions to the right and aligned to the left. A 16 pt non-serif font is used for large-print text and Marburg medium standard for braille.

descriptions and makes the map easier to carry, whereas the second allows for detaching legends after becoming familiar with the symbols or in the case of translating map legends into other languages.

Because of the high level of generalization on tactile maps, the information value of such maps is often much lower than of their visual counterparts. This is why it can be helpful to enhance tactile maps with audio content, either in the form of audio descriptions (or their textual equivalent) or direct audio feedback—for example, triggered when interacting with map features. Such content can be used to convey additional information that did not fit on the map. The combined use of audio and tactile information is especially helpful in presenting geographic data. Audio descriptions should be clear and concise without unnecessary clutter. A general map audio description should first state the purpose of the map and describe its layout, including a legend transcription for those not reading braille or not having residual vision. In describing the map content, audio feedback should take the form of a narrative that guides a reader through all its elements. This narration should begin from a characteristic place on a map—for example, the building entrance.

Standardization

A key principle of map symbology is to use standardized symbol sets whenever possible. However, tactile cartography lacks such standards because of perceptual differences among people with visual impairments, varying production techniques, and the limited extent to which a specific symbol can be used to denote a unique feature across a range of maps. It is not that researchers and practitioners have ignored this issue. The first attempts to create a standardized tactile symbol set were made in the 1970s and have been repeated many times since. Nevertheless, tactile cartography still lacks many conventions used in visual cartography (for example, green for forests), although some examples of local standardization exist. Some symbols are commonly used in particular regions of the world to denote specific features but can have entirely different meanings in other countries or regions.

Standardizing symbology, generalization parameters, and redaction rules could automate tactile map production, increasing accessibility by reducing production time and cost. Growing interest in inclusiveness has led to significant funding for research on accessibility and universal design. With initiatives supporting international and interdisciplinary cooperation (an example being the book you are reading), we hope that soon it will be possible to automatically generate legible and comfortable tactile maps of various scales and topics using open spatial data.

Conclusion

Some visual cartography design principles apply also to tactile maps. But tactile map design is not a simple transcription of visual maps into a raised form. This process should rather be understood as a translation of the spatial data into legible and meaningful products. Until now, however, most of the research has been focused on individual symbols, and relatively little attention has been paid to the more complex issues of evaluating how symbols can be combined most effectively to maximize legibility at the level of the whole map. Tactile map design research has employed a range of map production methods, partly dependent on the currently available technology, with relatively little attention to the suitability of these methods for different purposes.

Nevertheless, the solutions developed so far in the field of tactile cartography, considering the principles of universal design, can surely find wider applications, serving people with varying cognitive capabilities.

Finally, no matter how well designed a tactile map is, people with visual impairments must be aware of its existence for it to be used. Therefore, authors must not only design them well but also ensure sufficient exposure and dissemination.

Further reading

On design approaches in visual cartography

Bertin, Jacques. *Semiology of Graphics: Diagrams, Networks, Maps*, Esri Press, 2010.

Dent, Borden, Jeffrey Torguson, and Thomas Hodler. *Cartography: Thematic Map Design*, 6th ed. McGraw-Hill Education, 2008.

MacEachren, Alan M. *How Maps Work: Representation, Visualization, and Design*. Guilford, 2004.

Robinson, Arthur Howard, Joel L. Morrison, Phillip C. Muehrcke, A. Jon Kimerling, and Stephen C. Guptill. *Elements of Cartography*, 6th ed. John Wiley & Sons, 1995.

On the history of the cognitive turn in cartography

Keates, John S. *Understanding Maps*, 2nd ed. Routledge, 1996.

Montello, Daniel R. "Cognitive Map-Design Research in the Twentieth Century: Theoretical and Empirical Approaches." *Cartography and Geographic Information Science* 29, no. 3 (January 2002): 283–304.

Montello, Daniel R., Sara Irina Fabrikant, and Clare Davies. "Cognitive Perspectives on Cartography and Other Geographic Information Visualizations." In *Handbook of Behavioral and Cognitive Geography*. Edward Elgar Publishing, 2018.

On the cognitive tactualization approach and cognitive factors in tactile cartography

Blades, Mark, Simon Ungar, and Christopher Spencer. "Map Use by Adults with Visual Impairments." *The Professional Geographer* 51, no. 4 (November 1, 1999): 539–53.

Jehoel, Sandra, Don McCallum, Jonathan Rowell, and Simon Ungar. "An Empirical Approach on the Design of Tactile Maps and Diagrams: The Cognitive Tactualization Approach." *British Journal of Visual Impairment* 24, no. 2 (2006): 67–75.

Schiff, William, Emerson Foulke, Lester Krueger, Carl Sherrick, James Craig, David Warren, et al. *Tactual Perception: A Sourcebook*, edited by William Schiff and Emerson Foulke. Cambridge University Press, 2010.

On tactile map design guidelines and best practices

Antona, Margherita, and Constantine Stephanidis, eds. *Universal Access in Human-Computer Interaction: Design Methods and User Experience* 12768. Springer International Publishing (virtual event), 2021.

Braille Authority of North America. *Guidelines and Standards for Tactile Graphics*. Braille Authority, 2022. www.brailleauthority.org/sites/default/files/tg/Tactile%20Graphics%20Standards%20and%20Guidelines%202022_a11y.pdf.

Edman, Polly K. *Tactile Graphics*. American Foundation for the Blind, 1992.

ISO. *Accessible Design: Information Contents, Figuration, and Display Methods of Tactile Guide Maps* 19028. ISO, 2016.

NSW Tactual and Bold Print Mapping Committee. *A Guide for the Production of Tactual and Bold Print Maps*, 3rd ed. Sydney, 2006. https://printdisability.org/wp-content/uploads/2018/02/TabMap-Tactual-Maps-Guide-3-2006.pdf.

Wabiński, Jakub, Albina Mościcka, and Guillaume Touya. "Guidelines for Standardizing the Design of Tactile Maps: A Review of Research and Best Practice." *The Cartographic Journal* 59, no. 3 (2022): 239–58.

About the authors

Jakub Wabiński is assistant professor, Institute of Geospatial Engineering and Geodesy, Faculty of Civil Engineering and Geodesy, Military University of Technology, Warsaw, Poland, and is co-chair ICA Working Group on Inclusive Cartography. He received a BSc degree from the Maritime University of Szczecin in 2015 and an MSc from the Military University of Technology in Warsaw in 2017. He worked as a postgraduate researcher at the Dublin Institute of Technology in Ireland (2017) and a courtesy research assistant at the University of Oregon in the United States (2020). As part of his doctorate, he has been working on the issue of tactile map design and the automation of tactile map production. He is also interested in novel cartographic presentation methods. A board game enthusiast and a wanderer, he enjoys spending time in nature.

Jakub Wabiński.

Simon Ungar is a professional development tutor, Educational Psychology Group, University College London and an educational psychologist, Wandsworth Council, London. Simon earned his BA in psychology from the University of Sheffield in 1990, going on to complete a PhD in developmental psychology in 1994, looking at the question "Can blind and visually impaired children use tactile maps?" Following this, he gained funding for three years from the Economic and Social Research Council to continue research focusing on the design of tactile maps and graphics. He became a lecturer in developmental and environmental psychology in 1996 at Glasgow Caledonian University and later at University of Surrey. Having felt an increasing desire to apply psychology more directly in the community, Simon retrained as an educational psychologist in 2007, gaining his professional doctorate in child, family, and community psychology in 2010. He loves listening to music and urban walking, often at the same time.

Simon Ungar.

Notes

1. Peirce, Charles S. (1867), "On a New List of Categories," *Proceedings of the American Academy of Arts and Sciences* 7, 287–98.
2. Brooks, Joseph L., "Traditional and New Principles of Perceptual Grouping," in *The Oxford Handbook of Perceptual Organization*, edited by Johan Wagemans (Oxford Library of Psychology, 2015; online edn, Oxford Academic, March 3, 2014).
3. Montello, Daniel R., "Cognitive Map-Design Research in the Twentieth Century: Theoretical and Empirical Approaches," *Cartography and Geographic Information Science* 29, no. 3 (January 2002): 283–304.
4. Koláčný, Anton, "Cartographic Information: A Fundamental Concept and Term in Modern Cartography," *The Cartographic Journal* 6, no. 1 (1969): 47–49.
5. Jehoel, Sandra, Don McCallum, Jonathan Rowell, and Simon Ungar, "An Empirical Approach on the Design of Tactile Maps and Diagrams: The Cognitive Tactualization Approach," *British Journal of Visual Impairment* 24, no. 2 (2006): 67–75.
6. Wolfe, Jeremy M., "Guided Search 6.0: An Updated Model of Visual Search," *Psychonomic Bulletin & Review 2021* 28, no. 4 (February 5, 2021): 1060–92.
7. Brooks, Joseph L., *Traditional and New Principles of Perceptual Grouping*.
8. Blades, Mark, Simon Ungar, and Christopher Spencer, "Map Use by Adults with Visual Impairments," *The Professional Geographer* 51, no. 4 (1999): 539–53.
9. Aldrich, Frances, Linda Sheppard, and Yvonne Hindle, "First Steps Towards a Model of Tactile Graphicacy," *The Cartographic Journal* 40, no. 3 (2003): 283–87.

Chapter 6

Generalization for tactile maps

Guillaume Touya

In 2018, during a seminar at IGN, the French national mapping agency where I work, I met Jean-Marie, a researcher in computer science with a blind daughter. He'd come all the way from Clermont-Ferrand just to meet researchers in cartography. On his laptop, he showed me several pictures of tactile maps he'd recently encountered while accompanying his daughter. He asked me a simple question: "Can we take OpenStreetMap (the Wikipedia of cartography) and derive such maps automatically, on demand?"

Surprisingly, I had never thought about cartography for blind people and never imagined tactile maps as abstraction problems, but at that moment, it was obvious that map generalization was made for tactile maps. All my past research on map generalization (described later in the chapter) could find a brand-new outlet.

The more that we dived into the details of this problem, the more I began to connect tactile maps for people with visual impairments to past challenges of map generalization. You need space to sense the difference between close symbols? You can tweak displacement algorithms to separate those symbols. You cannot sense the details of the geometry of a line? You can push line-smoothing algorithms to more abstraction. I was at the same time fascinated by these new scientific challenges and motivated by the real need for my expertise.

A few weeks later, I convinced a student to start working with me on the use of automated map generalization in the production of tactile maps, and I never looked back. With this anecdote, I want to thank Jean-Marie for his friendship and for giving me the opportunity to dig into such fascinating questions.

—Guillaume

Content selection: Approaches

Comparing the content of tactile maps with visual topographic maps

Let's start with a simple comparison exercise: If we look at two versions of a map designed for the same goal, how exactly does the content differ? On the left, figure 6-1 shows a map of a park in Clermont-Ferrand, France, named Jardin Lecoq. The map was designed by a tactile graphics specialist, sometimes called transcriber—that is, a person whose job is to adapt different types of documents, including maps, for people with visual impairments. It is not just about making maps tangible but adapting them to the cognitive capabilities of a reader (see chapter 5, "Map Design and Cognition"). On the right, we show the same portion of space but from OpenStreetMap. Figure 6-2 summarizes the layers that are included in both maps or only in one.

Those two maps have the same goal but for different users: facilitate the pedestrian navigation within the park. As a result, the main layers, such as footpaths and buildings, or important obstacles, such as ponds and barriers, appear in both maps. But the semantic granularity in the tactile map is clearly reduced. Trees, flower beds, or visual points of interest appear only in OpenStreetMap. Stairs are also absent from the tactile map, and they do not even appear as footpaths on the map. In this way, the transcriber reduces the number of line symbols in the map and makes sure that the reader will not plan to use the stairs when traversing the park.

In this case, we cannot see any theme that is specific to the tactile map, but that is not always the case. For visually impaired pedestrians, two types of features usually do not appear in topographic maps: obstacles and landmarks that are specific to people with visual impairments[1] because they are based on senses other than vision.

Beyond the differences in terms of map layers, we can also note in figure 6-1 that some layers are not fully represented in the tactile map. For example, only one building is represented as a polygon, while another one north of the lake is represented as a braille toponym. The others are eliminated from the map during a selection process.

Selection processes for tactile maps

Map generalization can be seen as the cartographic equivalent of text summarization: Map generalization abstracts and simplifies geographic data to improve its rendering in a map. Selection is one of the main data transformations used during map

Generalization for tactile maps 107

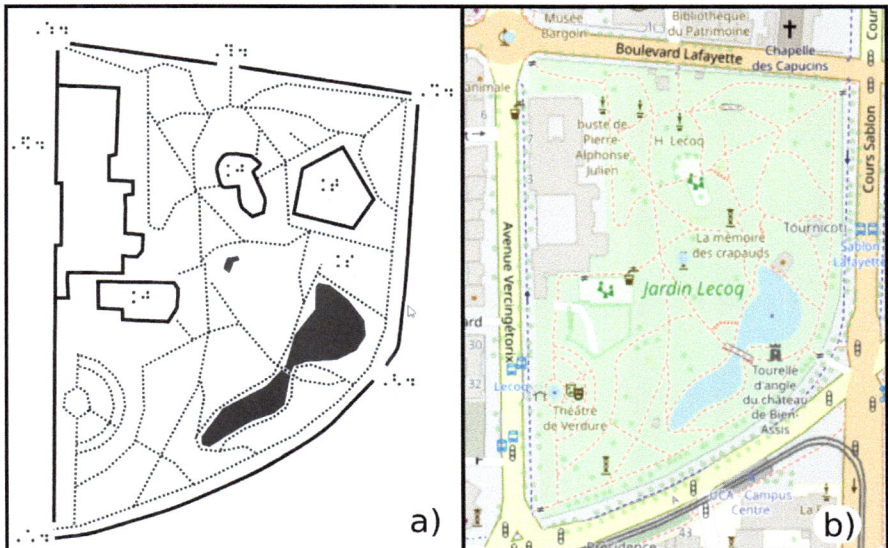

Figure 6-1. (a) A tactile map of Jardin Lecoq park in Clermont-Ferrand, France, and (b) its counterpart in OpenStreetMap. Image courtesy of Sophie Tricard; (b) ©OpenStreetMap contributors.

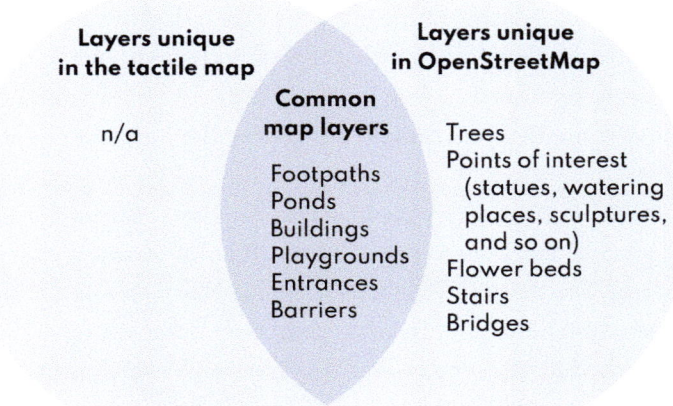

Figure 6-2. Common and unique layers in the two maps of Jardin Lecoq park.

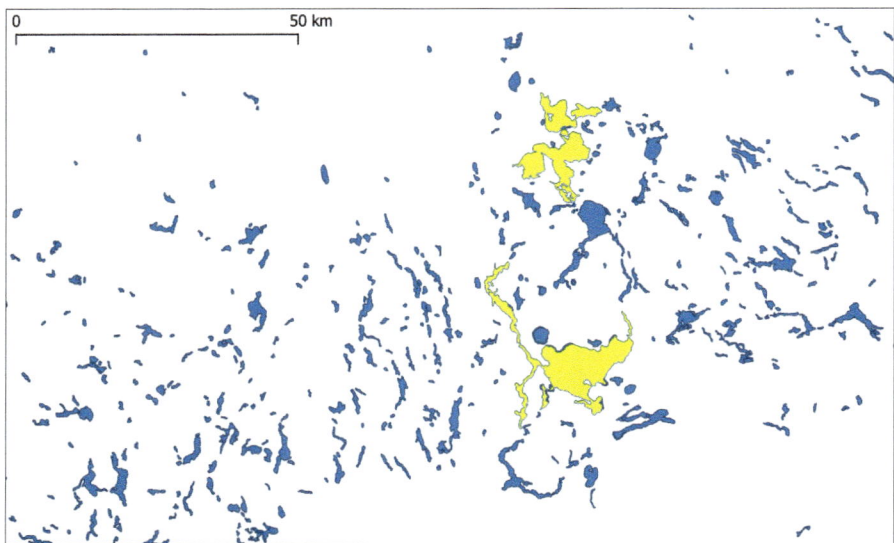

Figure 6-3. Selection of the most important lakes (with area over 100 km²) in Poland, during the generalization process for a tactile atlas of Poland.[2]

generalization. It is usually performed first, before simplifying the geometry of the map features. Sometimes called elimination, thinning, or pruning, selection determines which elements from the source geospatial dataset to keep in the map and which to remove. For example, figure 6-3 shows a selection process for Polish lakes during the design of thematic tactile maps of Poland. In this case, the selection is based on the area of the polygons: Only the lakes larger than 100 km² are selected and shown on the tactile map.

Most of the time, however, the selection process is more complex, for two main reasons:

1. We want to select and keep the important features in the map, but it is not obvious from the data precisely what "importance" means. A river may be important because of its length, its width in its final portions, its flow, or just because it flows through an important city (for instance, the Spree River in Berlin). Right now, researchers rely more and more on machine learning to solve this multiple-criteria decision task of selection.
2. The selection process should, as much as possible, preserve the main properties or characteristics of the space depicted by the map. When selecting roads in a map, a less important road might be selected only because it preserves the connectivity between two important roads. In the

case of figure 6-3, the selection based on lake area alone keeps only two big lakes in this specific region of Poland, characterized by a very high density of lakes. Consequently, the visually impaired map reader will not be able to read this information about lake density in this northern part of Poland. If the map designer wants to stress this lake density, a better selection would find a way to depict this specific geographic property while drastically reducing the number of lakes—for example, by clustering close lakes and aggregating each cluster into a larger lake.

These two considerations apply to any map creation process, not just tactile maps. Selecting the important features while preserving the main geographic properties of space is no different for a tactile map than for a visual map. What changes is how many features are selected and how many are eliminated, with a more drastic elimination when it comes to tactile maps. The case of toponyms for points of interest in figure 6-1 illustrates this need for more selection, because braille text, even with abbreviations, uses a lot of space on the tactile map. Only four toponyms remain inside the park in the tactile map, whereas the OpenStreetMap counterpart contains seven toponyms at that scale. Figure 6.1 is also a good illustration of how importance can vary in a tactile map: The statues and sculptures highlighted in the map for people with regular vision are not mentioned in the tactile map because this type of artwork is not generally accessible to people with visual impairments as landmarks to help orientation in the park.

Simplification of complex geographic data through generalization techniques

Pushing the existing generalization algorithms further

Even though visual and haptic perceptions cannot be directly compared, the "acuity" of haptic perception is lower than normal visual acuity (see chapters 3, "Understanding Through Touch," and 4, "Map Symbol Design: Visual and Haptic Variables"). The limits of visual acuity, from which minimal dimensions and then constraints are derived (figure 6-4), are the drivers of map generalization: We generalize maps when the represented features become too small or too close to each other to be visible on the map. Reduced acuity calls for larger minimal dimensions, and then stricter constraints.[3] Map generalization is usually difficult because the designer needs to achieve a balance between those constraints based on minimal dimensions and perception and the constraints to preserve, as much as possible, the map's information value. With tactile maps, this balance leans more toward minimal dimensions: the lines, polygons, and points represented in tactile maps should be more abstract, simple, smooth, and separated from one another than in their counterparts for people with normal vision.

The first consequence of these "bigger" minimal dimension constraints is that we can use the generalization algorithms designed for regular maps but with *pushed* parameters. When representing building polygons in a tactile map, we can use the algorithms from the literature that remove the small details (for instance, the small protrusion from the top building in figure 6-4), but using parameter values that

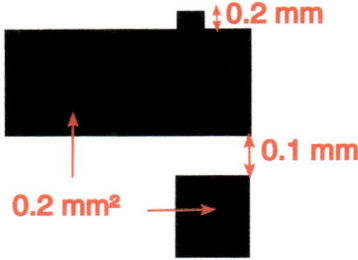

Figure 6-4. Classic minimal dimensions for buildings in a topographic map printed on paper. Those dimensions are the drivers of map generalization.

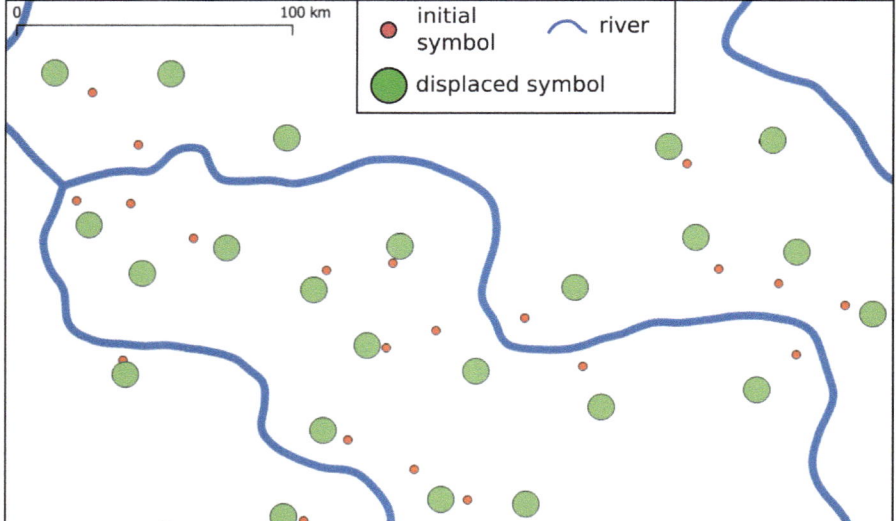

Figure 6-5. A large displacement of point symbols in a thematic map to allow enough room between symbols for haptic perception.

are considered too large for regular maps.⁴ When you want to simplify a line symbol, you can still use classical algorithms such as the Visvalingam-Whyatt algorithm, but with a larger area threshold to make sure that more vertices are removed from the line. Figure 6-5 shows another example in which a symbol displacement algorithm, initially designed to avoid collisions between symbols in a map, was successfully used to move symbols far away from one another and from river symbols in a tactile map.

More schematization and abstraction

Sometimes the transfer from visual to haptic perception requires more than just pushing the parameters of the algorithms; instead, we need more abstraction and more schematization. Schematized maps are maps in which the locational accuracy is sacrificed for a more abstract and direct understanding of the spatial relationships in the map. Metro maps are the most famous schematized maps, in which the location of the stations and the distance between two stations are more malleable than the connectivity between the lines and the general shape of the network. William Mackaness and Andreas Reimer define schematization in cartography as

> a process that uses cartographic generalization operators in such a way as to produce maps of a lower graphical complexity compared to maps of the same scale; the process aims to maximize task adequacy while minimizing nonfunctional detail.⁵

This definition applies particularly to generalization in the context of tactile maps, as task adequacy is much more important in tactile maps, which cannot offer a luxury of details. There is experimental evidence that schematized tactile maps are better understood than tactile maps that preserve more locational and distance accuracy.⁶

Schematization for tactile maps can take many forms. First, we can slightly rotate the map features when their orientation is close to 0° or 90° to make the perception of directions easier. This schematization principle was applied to roads in a large-scale map,⁷ inspired by the metro maps that usually allow only three orientations for the lines (0°, 45°, and 90°). Another way to apply schematization for tactile maps is to replace complex building polygons with simple geometric shapes, such as rectangles, *L*, or T-shapes. This schematization technique was used in the past for the generalization of topographic maps in regions such as Germany, the Netherlands, and Catalonia, where the small buildings were turned into squares. There are many cases of tactile maps in which buildings provide only the context of the

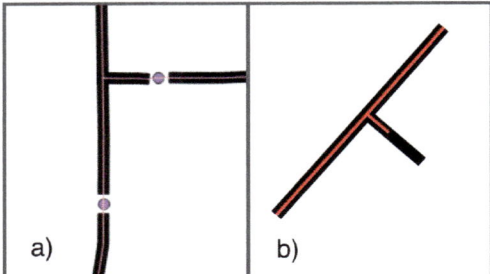

Figure 6-6. Two examples of generalization algorithms specifically designed for a tactile map: (a) the road symbol is cut around the crosswalk symbol; (b) the road is extended to be interpreted as a line when touching the map.

map—what Mackaness and Reimer call "nonfunctional detail"—and thus can be represented with a very abstract geometric representation.

Geometric abstraction can go even further by entirely removing any geometric detail beyond the location of a feature. For example, the selected lakes from figure 6-3 are simply replaced by a point symbol in a tactile atlas of thematic maps of Poland.[8] Similarly, one of the buildings of the park shown in figure 6-1, north of the large pond, is replaced by a braille abbreviation, which refers in the legend (not displayed here) to the name and the function of the building.

We also used different types of schematization algorithms when producing a tactile map describing a safe route for people with visual impairments on a campus in Nantes, France (figure 6-6). As crosswalks overlap road lines in the map, it is not easy to sense this overlap in terms of haptic perception. So, we exaggerate the spatial extent of the crosswalk and create a gap in the line around the crosswalk point symbol, to make sure readers sense the presence of a punctual symbol along the line (figure 6-6a). It is also not easy to sense the difference between a very short line and a point symbol. This is why we decide to exaggerate and extend the length of short lines (roads or sidewalks) to avoid confusion with a point symbol or a texture on the line (figure 6-6b).

Multiscale approaches to tactile cartography

Nowadays, people with normal vision have access to multiscale maps with applications such as Google or Apple Maps, where they can zoom in and out to visualize the world at multiple scales. Having dynamic access to multiple scales is extremely useful for many routine tasks with a map, such as route planning or wayfinding. Even though it may involve a bit of disorientation, users can relate to those different scales because the zoom is centered on the mouse cursor or the finger's location,

Generalization for tactile maps 113

Figure 6-7. Two sets of vector data created with TouchMapper, from which tactile maps can be printed, centered on the same address in Clermont-Ferrand, France, at two different scales, 1:1,400 and 1:3,200. ©OpenStreetMap contributors.

which gives a multiscale, multirepresentation view of space. People with visual impairments cannot access such tools, even though recent audio-vibratory prototypes are promising.[9]

The current solution to provide multiscale views of space to people with visual impairments remains the use of multiple tactile maps covering the same region at several scales. Several on-demand tactile map services, such as TouchMapper (figure 6-7), already provide maps at multiple scales, which can then be printed on different sheets of swell paper, or even on the same one if the map is small. As we can see with the maps in figure 6-7, the problem for visually impaired users can be to connect those two maps with their mental representations of space (see chapter 3). In the case of figure 6-7, the largest scale map contains mainly roads and buildings, but the smallest scale one does not contain buildings, so they cannot be used as landmarks to connect the two representations and to identify where the region covered by the first map is located in the second map. To make sure that the use of multiple tactile maps at different scales is feasible for people with visual impairments, map designers should provide several stable anchors in the maps. These anchors could be braille text or map symbols that are enhanced to be more salient with haptic perception. Map generalization techniques can be used to find the important features that can be used as anchors, and those anchors should then be simplified or abstracted carefully to ensure they are salient enough in all the scales.

Conclusion

We can assume that map generalization and tactile map design are a perfect match. On the one hand, map generalization is crucial to the production of tactile maps even more than for maps for people with normal vision because of the cognitive challenges placed on people with visual impairments. On the other hand, tactile maps are the ultimate display to explore the abstraction and simplification of geographic information. My advice for tactile map designers would be the following: Use map generalization as much as possible to make maps that are as simple and abstract as possible.

But we can also go beyond tactile maps. People with visual impairments often rely on vocal descriptions. These vocal descriptions also need some abstractions to avoid the long description of unnecessary details, and we believe that we could abstract space with map generalization before the generation of vocal descriptions. As map generalization tools and methods are focusing more and more on tactile maps, it will be easier for transcribers and cartographers to incorporate this key process in the design of tactile maps and thus make these maps more accessible for people with visual impairments.

Further reading

On research on map generalization
Burghardt, Dirk, Cécile Duchêne, and William A. Mackaness, eds. *Abstracting Geographic Information in a Data Rich World: Methodologies and Applications of Map Generalisation*. Lecture Notes in Geoinformation and Cartography. Springer, 2014.

Mackaness, William A., Anne Ruas, and L. Tiina Sarjakoski, eds. *Generalisation of Geographic Information: Cartographic Modelling and Applications*. Elsevier, 2007.

On practical tools to generalize maps
CartAGen library. https://cartagen.readthedocs.io/en/latest/.

Acknowledgments

Part of the work presented in this chapter was funded by the French National Research Agency, through the ACTIVmap project (ANR-19-CE19-0005).

About the author

Guillaume Touya is a senior researcher, IGN, ENSG, Univ. Gustave Eiffel, France, and is chair of the ICA Commission on Multi-scale Cartography. He holds a PhD in GI Science from Université Paris Est (2011). He has worked at IGN, the French national mapping agency, since 2004, and his interest in map generalization dates from his earliest mission at IGN. He is currently the principal investigator of the LostInZoom project (ERC grant no. 101003012). You can often find him listening to music on his earphones. He also enjoys reading novels and playing card games.

Guillaume Touya.

Notes

1. Wang, Min, Ni Zhu, Valérie Renaudin, Aurélie Dommes, and Myriam Servières, "Understanding and Using Spatial Landmarks of Visually Impaired People for Navigation Applications," in *2023 13th International Conference on Indoor Positioning and Indoor Navigation* (IPIN), 2023, 1–6.
2. Wabiński, Jakub, Guillaume Touya, and Albina Mościcka, "Semi-Automatic Development of Thematic Tactile Maps," *Cartography and Geographic Information Science* 49, no. 6 (August 31, 2022): 545–65.
3. Touya, Guillaume, Ridley Campbell, and Laura Wenclik, "Generalization Constraint Monitors to Assess Tactile Maps," in *Proceedings of 25th ICA Workshop on Map Generalization and Multiple Representation*, Delft, Netherlands, 2023.
4. Touya, Guillaume, Sidonie Christophe, Jean-Marie Favreau, and Amine Ben Rhaiem, "Automatic Derivation of On-Demand Tactile Maps for Visually Impaired People: First Experiments and Research Agenda," *International Journal of Cartography* 5, no. 1 (2019): 65–91.
5. Mackaness, William, and Andreas Reimer, "Generalization in the Context of Schematized Maps," in *Abstracting Geographic Information in a Data Rich World*, edited by Dirk Burghardt, Cécile Duchêne, and William Mackaness, Lecture Notes in Geoinformation and Cartography (Springer International Publishing, 2014), 299–328.

6   Wiedel, Joseph W., and Paul A. Groves, *Tactual Mapping: Design, Reproduction, Reading, and Interpretation* (US Department of Health, Education and Welfare, 1969).

7   Touya, Guillaume, Sidonie Christophe, Jean-Marie Favreau, and Amine Ben Rhaiem, "Automatic Derivation of On-Demand Tactile Maps for Visually Impaired People: First Experiments and Research Agenda," *International Journal of Cartography* 5, no. 1 (2019): 65–91.

8   Wabiński et al., "Semi-Automatic Development of Thematic Tactile Maps," 545–65.

9   Sargsyan, Elen, Bernard Oriola, Marcos Serrano, and Christophe Jouffrais, "Audio-Vibratory You-Are-Here Mobile Maps for People with Visual Impairments," in *Proc. ACM Hum.-Comput. Interact* 8, ISS, October 24, 2024, 551:624–48.

Case study

Tactile maps of historic gardens

Jakub Wabiński

Back in 2020, together with my colleagues from the Institute of Geospatial Engineering and Geodesy at the Military University of Technology, we asked people with visual impairments in Poland about the types of tourist attractions they usually visit. Most respondents indicated parks and gardens as the sites they visit most often. At the same time, they noted the lack of tactile graphics and maps that would enable them to fully explore such places. The results of this survey sparked an idea: to develop a technology to produce tactile maps of historic gardens, which are important elements of Poland's cultural and natural heritage. This led us to form an interdisciplinary team of cartographers, special pedagogues, biologists, and additive manufacturing experts to attain this goal.

In the words of Wojciech Malesa, a person with visual impairments and one of our project participants: "I was always striving to expand my knowledge. I was, and still am, a regular visitor to historical sites. I familiarized myself with every map, plan, or model that I could get my hands on. The problem is that there is not much of such material."

Most existing maps of parks and gardens are only a collection of green and blue patches depicting vegetation and water bodies traversed by alleys. However, what makes historic parks and gardens special are their unique compositions, from large lawns with scattered solitary trees in English gardens to the elaborately patterned garden parterres characteristic of baroque gardens. These compositions can be easily accessed by sighted participants, but people with visual impairments not only cannot see them but also, as is usually the case with historic sites, cannot touch them. Thus, the main purpose of our tactile maps was not orientation and mobility but rather a rendering of the characteristic features of parks designed in specific styles.

One of the study participants evaluating various printing techniques for tactile maps. Image courtesy of Jakub Wabiński.

Our main assumption for this project was that we should involve target user groups. We wanted to develop something applicable that would not remain a laboratory prototype. For this reason, during the three-year-long project, we planned six controlled study sessions with 20 people with visual impairments and diverse sociodemographic characteristics. The first two sessions were focused on symbology of the garden features, the next two were used to evaluate tactile map production techniques, and during the final two sessions, study participants evaluated our prototype maps. These sessions were supplemented by 100 hours of ad hoc consultations. All these direct meetings with our study participants provided us with invaluable feedback on the proposed solutions and, thanks to the use of rapid prototyping techniques, allowed us to tweak them iteratively.

Using the technology we developed, we created five pilot sets of tactile maps, using the ultraviolet (UV) printing technique—a type of additive manufacturing technology, in which full-color light-curing resin is sprayed onto a substrate and immediately solidified by UV light. Each of the pilot tactile maps consists of two or three maps at different scales and levels of detail, for the Terrace Gardens in Książ, the Baroque Garden in Wilanów, the Romantic Park in Arkadia, the English Park in Krasiczyn, and the Japanese Garden in Wrocław.

"I have visited the Baroque Garden in Wilanów many times before," reported Wojciech Malesa, "but I was not aware of how intricately decorated the garden

One of the quarters in Terrace Gardens in Książ, consisting of symmetrical flower beds surrounded by hedges, with a fountain in its central part and single containers with ornamental shrubs in each corner. From left to right: photograph of an original element, generalized vector drawing of the tactile layer, and UV printed element in full color. Image courtesy of Jakub Wabiński.

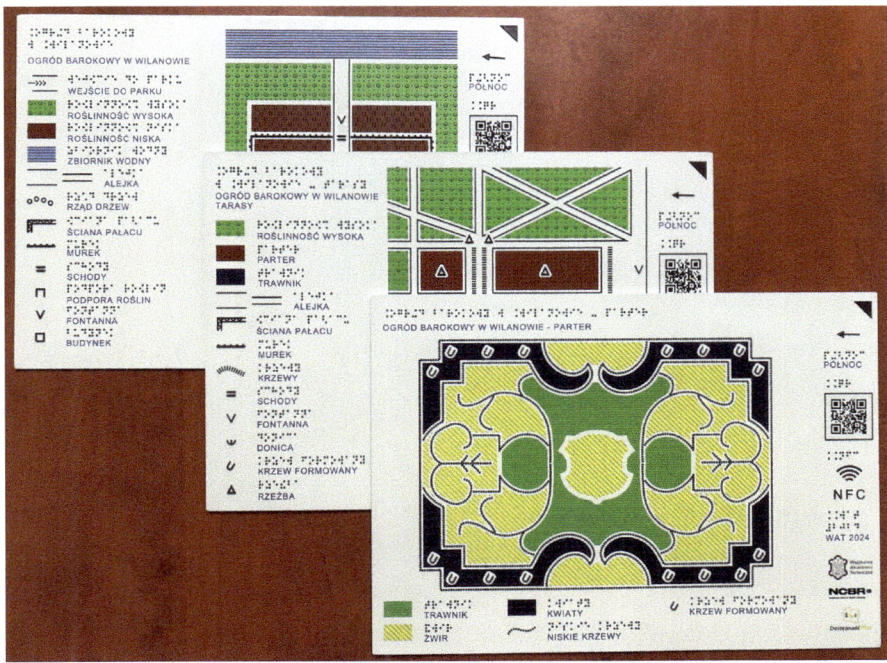

A set of three UV printed tactile maps showing the Baroque Garden in Wilanów at three levels of detail. The first map presents the whole garden composition and the second one shows one of the terraces, whereas the large-scale map depicts one of the rectangular garden parterres. Image courtesy of Jakub Wabiński.

parterres were in reality. Thanks to these tactile maps, I could finally feel all the details, and now I fully appreciate the ingenuity and artistic sense of this garden's designers."

These maps are supplemented with audio descriptions, which can be accessed using a QR code or NFC tag, as well as descriptions in braille and in large print. The maps will be used by tourists visiting the parks and by centers for education of people with visual impairments, as well as by members and beneficiaries of the Polish Association of the Blind.

Finally, the usefulness of our maps was verified during a field trip to one of the parks, during which children with visual impairments and their guardians had an opportunity to feel the magical atmosphere of the romantic park through different senses. The smiles on their faces were the best reward for our team!

One of our maps scanned during the field trip. Image courtesy of Jakub Wabiński.

Case study

Making the invisible visible

Shirly Goldner

The State of Israel has a population of about 150,000 people with visual impairments who navigate their lives relying heavily on touch, sound, memory, and assistance from others. For most of us, using a smartphone or unfolding a map to find our way is second nature. But imagine trying to understand your surroundings without ever seeing them. This is the reality for Israel's visually impaired community, who until recently had no access to large-scale tactile city maps.

The story began when the Survey of Israel (SOI) recognized this critical gap in accessibility. The SOI saw how the lack of tactile maps created a substantial "blind spot" in spatial understanding for this community. Although technology provides voice navigation and audio descriptions, these tools lack something fundamental: orientation, the ability to understand one's surroundings through touch and other senses—a skill the blind rely on profoundly. Both the blind and people with residual vision require simple, legible maps to comprehend their surroundings.

This insight led to an ambitious social project: creating free, custom-made tactile maps for people with visual impairments. The SOI collaborated with Kadaster-Netherlands, the Center for Accessible Culture, the Ministry of Education, the Center for the Blind in Israel, and, importantly, working groups of people with visual impairments who became both consultants and end users. "Every journey requires careful planning when you're blind," said one project participant. "Audio helps, but it's like trying to understand a painting through someone else's description. You need to feel it to truly grasp it."

Visual maps often suffer from "information overload": They are densely packed with details, colors, symbols, and text, making them impossible for people with visual impairments to use. Also, existing maps lacked braille and raised symbols, making them inaccessible to people with visual impairments.

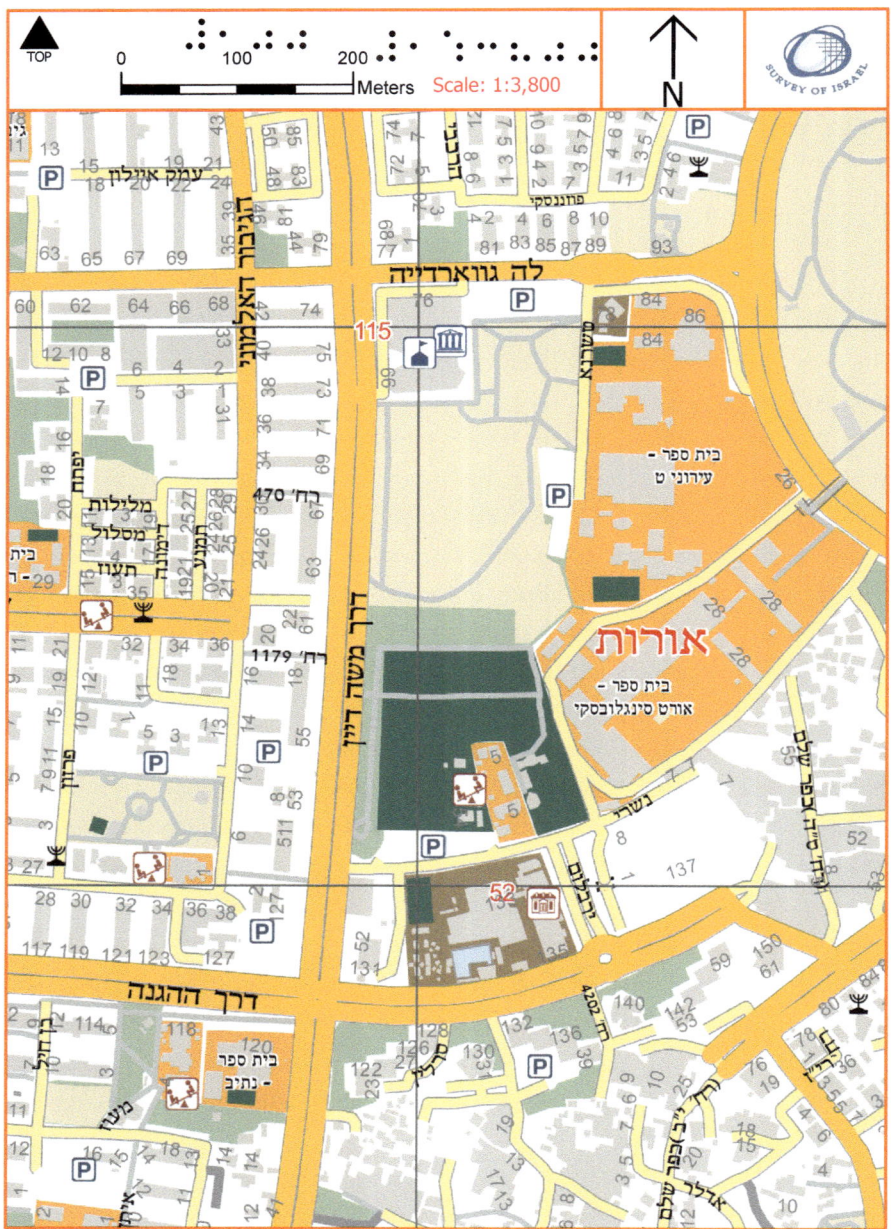

Traditional map – Center for Accessible Culture Area. Image courtesy of Survey of Israel.

To solve this problem, the SOI embraced the principle of purposeful simplification. Instead of trying to include all information on a single map, the SOI separated themed maps for specific purposes, each encompassing essential features only. Examples of themes include city maps showing only main streets, landmarks, bus stops, and train stations; tourist maps with cultural sites, benches, and amenities; and state and world maps including borders and seas. For its first study, the SOI researched large-scale city maps, for understanding the surroundings for navigation purposes.

Each element on the map served a specific purpose, and only main streets were included, along with locations important to each user. Crosswalks with and without traffic lights were marked distinctly. Only simple symbols were used, while points of interest were coded with braille numbers to reduce the number of symbols. Minimizing unnecessary features and differentiating themes makes the map clearer and allows users to focus on essential details of the subject at hand.

Before using the symbols, the SOI conducted research to determine which symbols, text, and colors were comprehensible for people with visual impairments. Symbols were limited to simple shapes, such as lines, dots, circles, triangles, and squares, spaced 5mm apart to avoid tactile confusion.

The SOI sought to create cost-effective maps to make them widely accessible. Tactile maps in swell paper technique proved ideal because they are relatively inexpensive compared with 3D-printed options and can be easily distributed. The SOI used Swell Touch paper, which creates raised surfaces on black ink through heating. Each map fits on A3 or A4 paper, making it lightweight and portable. For people with residual vision, digital maps were a great solution because of the contrast of colors that can be seen through the computer and phone.

Although colors are invisible to the blind, they were found to be crucial for other people with visual impairments, who benefit from high-contrast visuals. Symbols to be extruded were printed in black for swell print recognition, whereas contrasting colors were used for other features, enabling the visually impaired to distinguish them.

For text, standard Arial font was used, whereas braille was minimized and fully spelled out in a separate legend sheet that accompanies the map. The map contains writing in contrast colors for the visually impaired and for the blind person's guide.

In conclusion, the SOI discovered that creating tactile maps was not just a cartographic challenge but an exercise in understanding a different perspective in comprehending maps. Each individual experiences these maps differently, and what

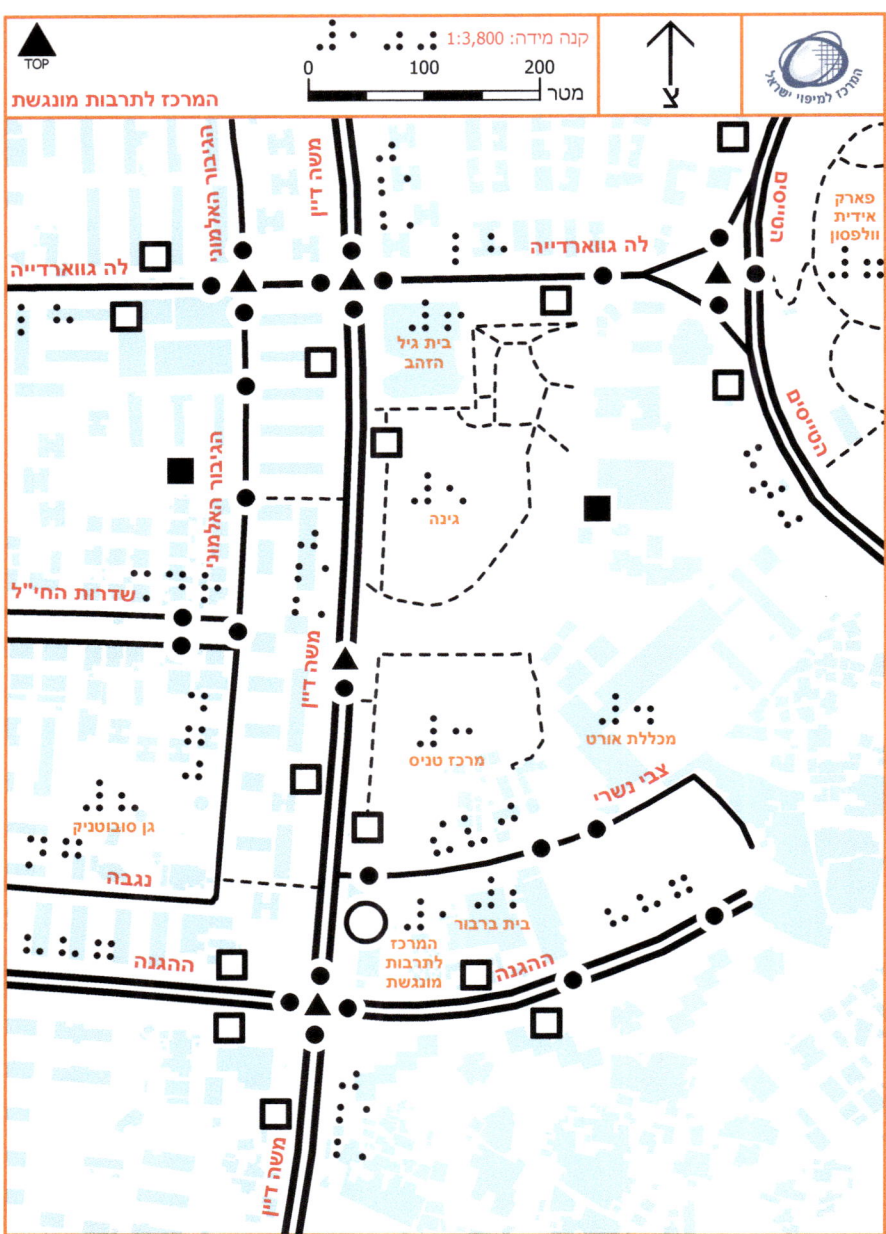

Tactile map – Center for Accessible Culture Area contains black symbols meant to be raised to a tactile form, as well as labels in both standard large print and braille. The map sheet includes a title, orientation mark, and north arrow, as well as a linear scale and the Survey of Israel logo. Image courtesy of Survey of Israel.

works for one user might not suit another. This understanding led to the SOI's flexible approach of customizing maps by theme and minimizing features, using clear and simple symbols based on each user's individual needs.

Today, SOI continues to produce maps, adhering to core guidelines. Although users require some initial planning and guidance, these maps are opening new possibilities for independence and spatial understanding in Israel's community of people with visual impairments. This initiative highlights how thoughtful design can break down barriers, creating tools that enhance lives. It's a reminder that meaningful innovations sometimes come not from adding features but from choosing carefully what to leave out—ensuring that what remains truly serves its purpose.

Tactile map legend. Image courtesy of Survey of Israel.

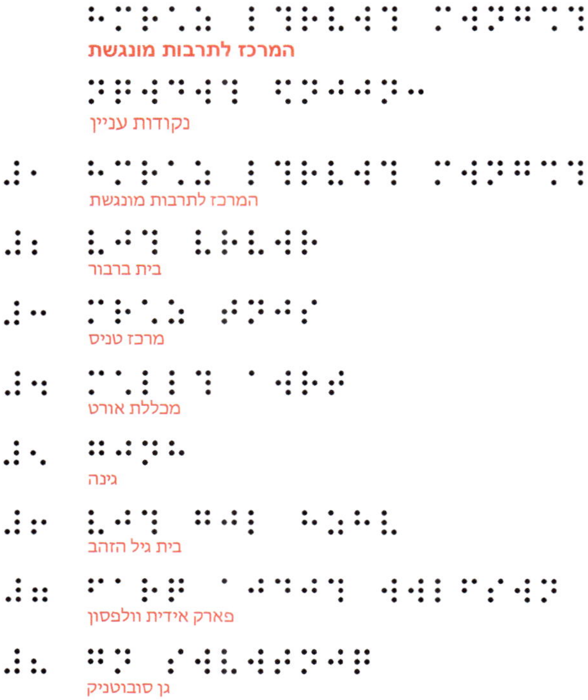

⠓⠍⠗⠅⠌⠀⠇⠮⠗⠃⠧⠎⠀⠍⠯⠝⠣⠱⠮

המרכז לתרבות מונגשת

⠝⠴⠬⠊⠙⠎⠀⠑⠝⠊⠊⠝⠼

נקודות עניין

⠼⠁⠄ ⠓⠍⠗⠅⠌ ⠇⠮⠗⠃⠧⠎ ⠍⠯⠝⠣⠱⠮

המרכז לתרבות מונגשת

⠼⠃⠄ ⠃⠊⠮ ⠃⠗⠃⠧⠗

בית ברבור

⠼⠉⠄ ⠍⠗⠅⠌ ⠉⠝⠊⠎

מרכז טניס

⠼⠙⠄ ⠍⠅⠇⠇⠮ ⠁⠧⠗⠉

מכללת אורט

⠼⠓⠄ ⠣⠊⠝⠓

גינה

⠼⠧⠄ ⠃⠊⠮ ⠣⠊⠇ ⠓⠱⠓⠃

בית גיל הזהב

⠼⠌⠄ ⠋⠁⠗⠅ ⠁⠊⠙⠊⠮ ⠧⠇⠋⠎⠧⠝

פארק אידית וולפסון

⠼⠚⠄ ⠣⠝ ⠎⠧⠃⠧⠉⠝⠊⠅

גן סובוטניק

Tactile map legend and code sheet. Both large-print and braille text are applied to the code sheet. Image courtesy of Survey of Israel.

Part IV
Users and education

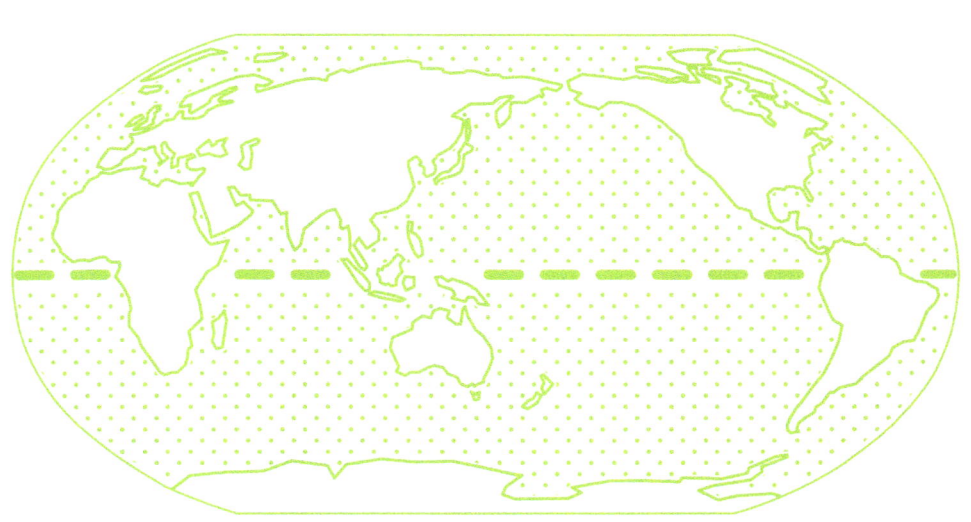

Personal story

Completely lost without maps

Leydiane Cristina Santana

Leydiane holding a map of Brazil.

When my mother was pregnant with me, she got infected with toxoplasmosis, which caused me to be born with low vision. Initially, my sight was about 20 percent. But at the age of seven, I lost my sight entirely. Now at 35, I no longer even perceive light.

In Brazil, being a student with a vision impairment was very difficult. Until the age of seven, I learned to read using enlarged letters. But after that, I had to learn braille, take notes in class using a *reglete*, a ruler with holes in it that helps users write braille, and do calculations with a soroban, an abacus for people with visual impairments.

None of the materials at school were adapted for people with low to no vision. Teachers delivered lessons to the whole class, without describing images or explaining certain concepts in greater detail. Only my chemistry teacher tried to adapt some materials. In geography, there were no tactile maps or any other accessible resources, so I had to follow along solely by listening to the teacher's explanations.

The first time I touched a tactile map was at São Paulo State University. I took part in a preparatory course that helps students study for the Vestibular, a popular entrance exam for Brazilian universities. One of the volunteer teachers introduced me to a tactile map of the regions of Brazil, followed by a map of Brazilian biomes.

They were amazing! Before, I only knew about Brazil's regions from hearing about them on TV, and I had no idea how they were shaped.

With help from each map's legend, I could understand the regions and biomes. All tactile maps should include legends! The textures of the maps were also fantastic because they allowed me to distinguish each biome. I loved the biomes map so much that I asked my teacher if I could take it home.

I have now also interacted with an adapted globe. To teach me about parallels and meridians, São Paulo State University geography professor Carla Sena ran textured ribbons around the globe to reveal the Tropic of Capricorn and the Tropic of Cancer. This helped me learn about the seasons in different hemispheres.

Without maps, students with visual impairments can get completely lost. Just listening to an explanation isn't enough to learn geography. I hope that all maps and globes can be adapted to better support geography education for people with visual impairments.

Chapter 7

User-centered and inclusive cartographic design

Robert Roth, Merve Keskin, and Zdeněk Stachoň

I knew what I wanted to do with my life by the second week of my introductory cartography class. In my first year of college, I declared as a mathematics major (which became a bit of a bore for me because too often there was only one answer to the problem), but in high school I had taken every available art class, including more tactile forms, such as pottery, embossing, and printmaking. Cartography is the perfect blending of both data science and creative art; I fully tricked my family into thinking I am an engineer when I know I am a designer! To my surprise, however, I learned later in that same cartography class that I have what was labeled as a major disadvantage to becoming a professional cartographer: I have red-green color vision deficiency and thus cannot use many conventional maps that rely on a "stoplight" color metaphor or that place warm hues atop a green or brown satellite basemap. However, what I feared would be a barrier became a critical early lesson, as it taught me the benefit of designing from the margin, as well as to be aware of different, far greater needs of other people with visual impairments and to include these users throughout the design process. Today, I direct the University of Wisconsin–Madison Cartography Lab, and we seek to make our work as widely accessible as possible. However, our primarily visual work in the Cart Lab, and in many cartographic institutions, barely meets the basic needs for people with visual impairments. I see tactile mapping, along with other forms of visual and nonvisual accessibility, as one of the greatest challenges facing cartography today, and one that requires participation and ownership from people with visual impairments to move forward.

—Rob

In 2015, I visited "Dialogue in the Dark" in Istanbul, Türkiye, an immersive experience where blind guides lead visitors through completely darkened spaces, simulating everyday environments, such as a park, a café, or even a boat ride. As a young cartographer back then, I thought I understood spatial awareness, but navigating without sight was transformative. Simple actions such as finding a seat or identifying a texture became puzzles that required new strategies. This experience redefined my understanding of how nonvisual senses shape spatial perception. I realized the profound role tactile maps play for people with visual impairments not as simplified representations of space, but as gateways to exploring and understanding the world in ways that are often invisible to sighted individuals. This lesson has stayed with me, inspiring my approach to creating inclusive cartographic designs.

—Merve

I have been fascinated by maps since childhood, captivated by how they allow us to study places we never have seen in person. During my studies, I discovered a passion for cartography and decided that maps would be my lifelong journey. As part of my studies, I had the opportunity to visit The Support Centre for Students with Special Needs at Masaryk University in Brno, Czech Republic, where I encountered tactile maps designed for haptic perception. I was amazed by how people with limited or no visual perception could create and use spatial representations. The more I learned about these maps, the more one fact surprised me: The very people for whom these maps were designed were rarely involved in their creation. I asked, "How can maps truly serve their purpose if their users are not part of the process?" I have since searched for answers directly from people with visual impairments.

—Zdeněk

Introduction: Our user experiences

Together, we form the core leadership group for the International Cartographic Association (ICA) Commission on the User Experience, where we advocate for designing mapping products with direct input from their intended users. In this chapter, we use the term "mapping products" to describe the software, hardware, and materials available for designing maps as well as the resulting maps, tactile or otherwise. We adopt arguments from user-centered and inclusive design that recruiting users who are often found at the margins of design is not only ethical because it supports universal human rights but also results in a more robust user experience for all. We first introduce key tenets of user-centered and inclusive design and then summarize a six-stage user-centered design process we employ in our work. Next, we summarize design recommendations about tactile mapping derived from user studies. We conclude by adopting arguments from disability and design justice that the design process should be led and controlled by the users who are most affected by the resulting designs.

Problem: User-centered versus inclusive cartographic design

User-centered design describes an early and active focus on user needs when planning and implementing a product—map or otherwise—with an emphasis on iterative participation and refinement during the design process (after Norman 1988). A product is rarely made from an idealized design cycle where the designer has a design idea and implements it perfectly on the first attempt to an overwhelmingly positive response (figure 7-1a). Instead, the creative design process has been conceptualized as a meandering "squiggle," in which the designer first considers numerous alternatives, eventually lands on one or several working prototypes, and ultimately executes the final design (after Newman 2009; figure 7-2). This standard design cycle centers the designer and thus the designer's needs, interests, and values. Although sometimes necessary and successful, such a standard design cycle runs the risk of designing a perfectly functioning product that meets the needs of no actual users, requiring an endless cycle of revisions and extensions to find the actual user audience (figure 7-1b).

In contrast, a user-centered design cycle relies on input and feedback to navigate this "squiggle," centering the process on intended users and their needs, interests, and values. Thus, a user-centered design plans for a series of user evaluations to

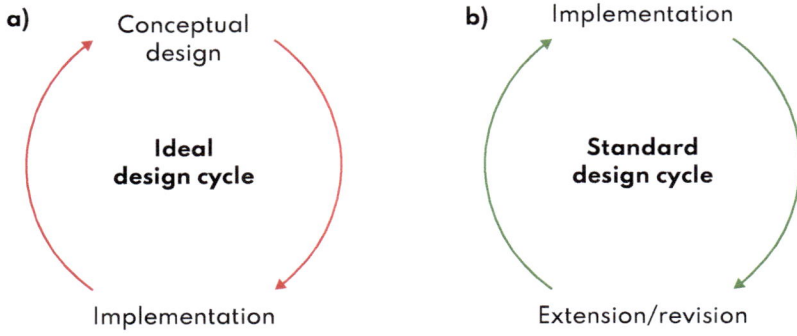

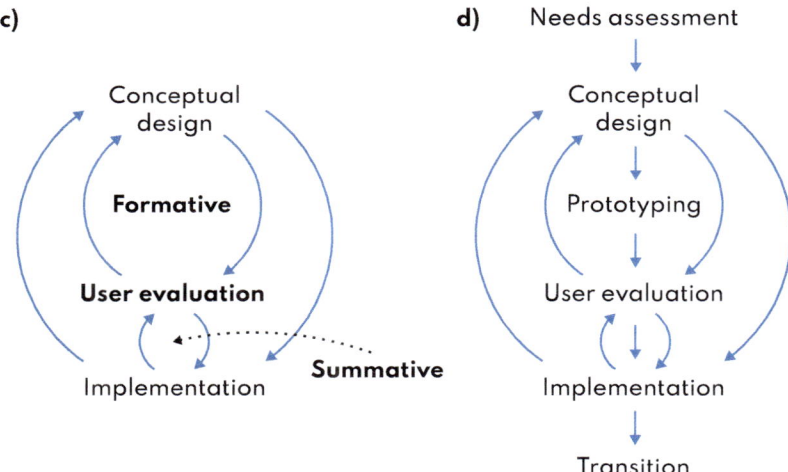

Figure 7-1. Alternative Design Processes: (a) The idealized design cycle seeks a single progression from conceptual design to implementation. (b) The standard design cycle instead requires multiple iterations from implementation to extension/revision during design, running the risk of designing a product that meets no real needs and ultimately exhausting project resources. (c) The user-centered design cycle uses user evaluation to mediate iterations between conceptual design and implementation, calibrating the design to real user needs through user input and feedback during the process. Early formative evaluation should be prioritized over late summative evaluation. (d) Our six-stage user-centered design approach that includes needs assessment, conceptual design, prototyping, user evaluation, implementation, and transition. Figures adapted from Roth, Ross, and MacEachren 2015; Roth 2019.

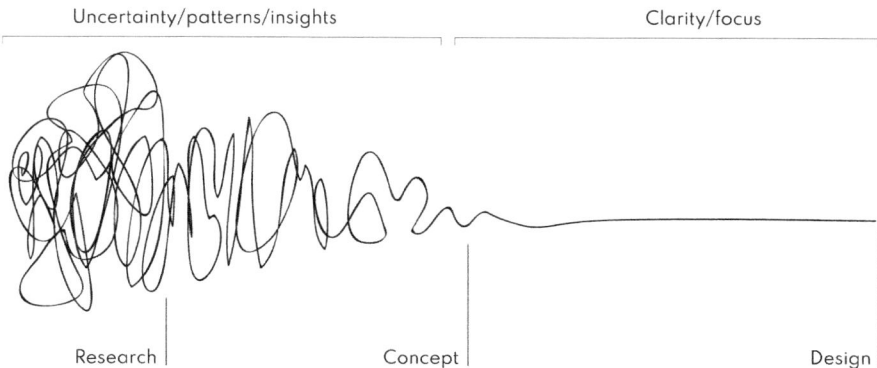

Figure 7-2. The Design Squiggle. The "design squiggle" is a metaphoric representation of the creative design process. The line itself represents time, with most creative time spent during research and prototyping. Early in the project, uncertainty is high as a range of alternatives are considered. Input and feedback from users can help delineate ideas that support real needs versus less useful alternatives and suggest additional ideas not considered by designers. Clarity and focus are achieved toward the end of this creative effort, resulting in streamlined implementation of the product. If users participate in the early creative stage, this streamlined final implementation is more likely to meet real user needs and be used for its intended purpose. Figure republished from Newman 2009 following CC-BY-ND license.

bridge the gap between the design concepts and their implementation (figure 7-1c). Although some project managers may hesitate to allocate resources to user feedback, research shows that input from just three to five target users can uncover up to 80 percent of potential design issues in each evaluation session (Nielsen 1992).

Despite its benefits, a user-centered approach has been critiqued from both science and industry for too narrowly defining the target user. In science, user studies are explicitly aligned with the central limit theorem, which presupposes that the distribution of an empirical sample will conform to a normal bell curve as the sample size grows (figure 7-3a). Accordingly, generalizable design insights—including many empirically derived cartographic design conventions—are formed for the mean or average user in the sample. Users at the margins are peripheralized as outliers and ignored as statistically improbable.

In industry, user-centered design is implicitly aligned with Rogers's (1962) "adoption curve," which co-opts the bell curve to define a timeline for user adoption (figure 7-3b). Notably, profitable design insights are formed by overcoming the chasm of early adopters to reach the early and late majority. Users at the margins

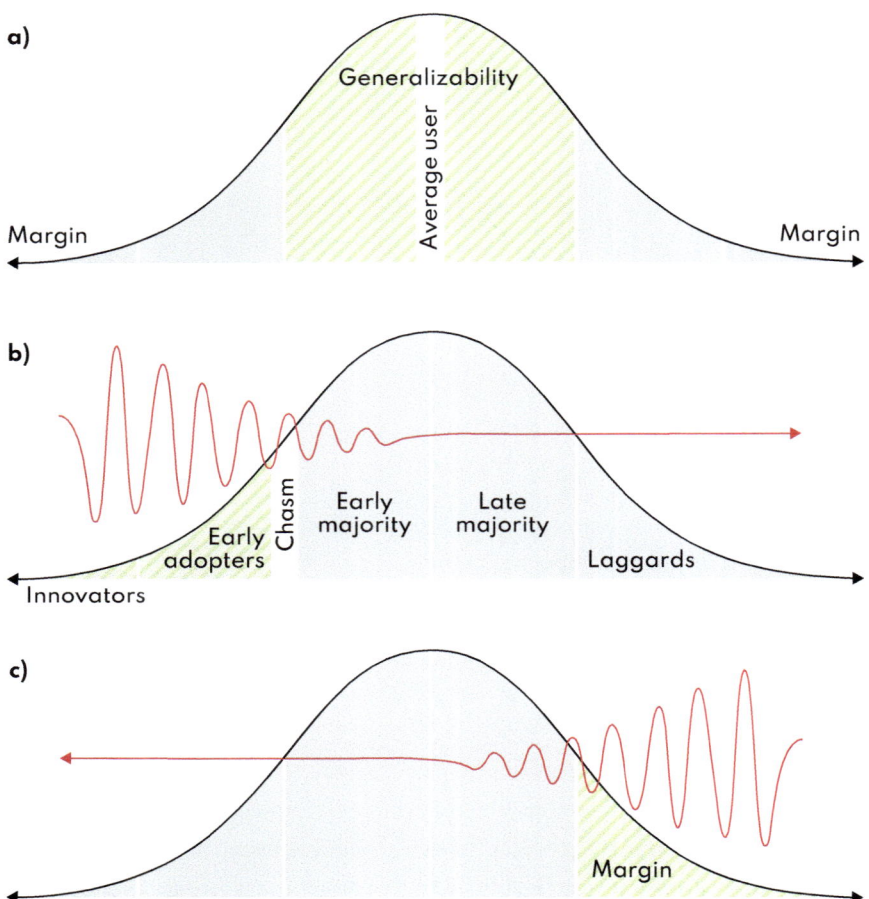

Figure 7-3. User-centered versus Inclusive Cartographic Design: (a) The bell curve describes the distribution of an empirical sample under conditions of normality, with the mean in the center, 68 percent of observations within standard deviation of the mean, and 95 percent of observations within two standard deviations of the mean. Scientific research on human subjects seeks generalizable insights that apply to the "average" user in the middle of the bell curve, leading to suboptimal designs for users on the margin. (b) Rogers's adoption curve suggests that the diffusion of innovation starts with innovators and early adopters, representing a small percentage of potential users. Often, a large amount of effort is spent on these highly motivated user groups, as indicated by superimposition of the "design squiggle" on the curve. Industry profitability is reached in overcoming the "chasm," with the majority adopting the innovation. Laggards are left to the margin. (c) Inclusive design begins with users on the margin, effectively flipping the "design squiggle" to spend the most effort on the user groups often missed by practices in science and technology. Spending time and resources at the margin leads to a more robust experience for all. Figures adapted from Rogers 1962.

again are peripheralized as "laggards" and considered unlikely to contribute to the consumer market.

Inclusive design addresses the gap in traditional user-centered design by starting not with the average or majority, but with users at the margins (after D'Ignazio and Klein 2020), emphasizing accessibility, disability, and justice. Conceptually, inclusive design inverts the design squiggle, starting the process from the most constrained user experience (figure 7-3c). Yet designing for the margins is not only ethical but also results in an improved user experience for all (Assembly 1949), because everyone temporarily or situationally loses abilities (Microsoft 2019).

Many intersecting factors can push users to the margins of a design, including gender, sex, sexual orientation, race, ethnicity, nationality, geography, language, religion, age, education, and socioeconomic status, among others. Variations in perceptual, cognitive, physical, and emotional abilities create further marginalization, particularly for those facing permanent, temporary, or situational disabilities. People with visual impairments, in particular, have been historically underserved in cartography, often excluded from user-centered design processes. This exclusion perpetuates a visual bias in mapping products.

The principles of user-centered and inclusive cartographic design outlined here are crucial for any tactile mapping project. However, the user-centered and inclusive design principles described here are relevant to all mapping projects, not just tactile maps, as all projects benefit by including tactile or other nonvisual alternatives.

Process: Enabling user participation during design

User participation from the start of a project offers significant advantages over seeking user input only at the end of the design process. Effective user-centered design formalizes this early engagement through structured, iterative stages, each with clear milestones. This approach ensures meaningful and consistent feedback from users throughout the process. Many user-centered frameworks, including those specific to cartography, emphasize the importance of the distinction between formative and summative evaluation (figure 7-1a).

Formative evaluation involves gathering user input early in the design process, starting with feedback on user needs before design begins. By planning for formative participation from the outset, designers can incorporate user needs and refine the product as it evolves. In contrast, *summative evaluation*, conducted after the product is nearly or fully complete, offers limited actionable insights. At best, it may inform future projects; at worst, it can highlight fundamental design flaws that require complete redesign, undermining the entire effort. Prioritizing early and iterative user involvement prevents such costly missteps and ensures that the final product effectively meets user needs.

In practice, we recommend a six-stage user-centered design process that has, by and large, stood the test of time in our research and design projects (figure 7-1d; based on Robinson et al. 2005; updated terminology from Roth, Ross, and MacEachren 2015).

1. **Needs assessment:** A *needs assessment* is a textual description of user needs, interests, and values based on user participation. Major milestones from the needs assessment include articulation of user personas and use case scenarios. *User personas* are background descriptions of several anonymized or hypothetical users based on conversations with actual intended users. *Use case scenarios* are narrative descriptions of how users will use the mapping product, as well as for what purposes. These written descriptions of users and use cases help humanize the design process for designers, ensuring that they keep real user needs in mind rather than drifting toward their own interests and values. User personas provide a concrete starting point for users to voice concerns about the purpose of the mapping product and for whom it is and is not designed. Following inclusive design, the needs assessment should identify user personas that explicitly vary by intersectional user differences, including disability type, to begin the design process from the most constrained use cases. The needs assessment also should include sometimes difficult discussions between stakeholders and users about ownership over the data and designs generated from the project, as well as strategies for long-term maintenance to ensure user access over time. In this way, the needs assessment stage establishes a strong relationship between designers and users, with the goal of understanding the rich use and user context informing a new project. It also helps avoid decontextualized "parachute" design that imposes existing tools and ideas onto participating users.

2. **Conceptual design:** After processing user input from the needs assessment, the designer completes their first *conceptual design* or a textual description of the functional and nonfunctional scope of the project. The primary milestone during conceptual design is a *requirements document* that outlines

all datasets, maps, and any additional interactive, tactile, or multisensory features required for the mapping product. The requirements document also includes a list of nonfunctional contextual considerations for the design, such as constraints placed on design by the tactile medium, production technology, or map-use environment, as well as other considerations regarding accessibility, disability, and justice. In user-centered design, the set of requirements is sometimes described as the design *utility* or the overall usefulness of the mapping product for intended users. Inclusive design suggests that not all product utility will be equal for all users, but considering the needs of users at the margins ensures that the mapping product will remain useful to all users should they temporarily or situationally lose abilities. The conceptual design stage can also include a *comparative analysis* of similar or related mapping products to understand potential gaps in utility that could set the new product apart from others, particularly for users at the margins.

3. **Prototyping:** Compared with the textual conceptual design, *prototyping* creates a visual or, for people with visual impairments, audio and/or tactile description of the functional and nonfunctional scope of the project. Major milestones from prototyping include a *wireframe*, or a rough prototype outlining the planned map elements and their layout, and a *specification sheet*, providing examples of symbolization, labeling, and other styling design decisions. For tactile mapping, considering the costs of production, we recommend high-fidelity (hi-fi) wireframes that use example data over low-fidelity (lo-fi) wireframes using placeholder text or meaningless maps. Prototyping also can be supported by 3D printing technology that may not be of sufficient tactile quality for final production but offers a fast and affordable medium to support rapid prototyping. In user-centered design, prototypes help intended users comment on the *usability* or ease of using a mapping product relative to the utility outlined in the requirements document. Again, inclusive design suggests that not all product usability will be equal for all users, but considering efficiency and error frequency for users at the margins will improve the likelihood that the mapping product will remain usable for all users if their needs change.

4. **User evaluation:** A *user evaluation* is a test of a version of the mapping product and emphasizes the iterative nature of the user-centered design process. A user evaluation can be on the needs assessment, the requirements

document, or prototypes, but, importantly, each user evaluation initiates a round of revision to these milestones that can be conceptualized as a *user → utility → usability loop* (after Roth, Ross, and MacEachren 2015). Emphasizing formative over summative user evaluation typically results in fewer loops and thus more quickly arrives at a codesigned mapping product that works for a diverse range of possible users. As introduced above, user evaluations can include just three to five users with each loop, if the participating users are representative of wide intersectional user differences (Nielsen 1992). For tactile maps, we recommend in-person interviews, focus groups, and cognitive walkthroughs (that is, asking the user to "talk" or "think" aloud while using the prototype). It is important for people with visual impairments to use the physical prototypes—which require additional accommodations and considerations for co-located participation scheduling and testing procedures—rather than online surveys or other distributed evaluation methods. User evaluations of tactile maps should prioritize users with visual impairments. Evaluation with sighted individuals who are blindfolded, placed in a darkened room, or whose vision is otherwise temporarily suspended is not a replacement for input and feedback from people with visual impairments.

5. **Implementation:** *Implementation* describes the actual execution of the mapping product design. User-centered and inclusive design can mean that the designer instructs intended users how to make their own maps or mapping products, and there are some design contexts where migrating complete design ownership to users is necessary. However, such participatory codesign more commonly requires designers to draw on their design expertise for the implementation, without assuming expertise on the map purpose and content itself. For user-led design with designer-led implementation, it is important to maximize input and feedback from people with visual impairments while minimizing their time on the project. Depending on the scope of the mapping product, there may be a single implementation stage after one user → utility → usability loop or a schedule for several milestones, such as an *alpha* (v0.1; partially featured), *beta* (v0.2; fully featured but with errors), and *final release* (v1.0; fully featured, debugged, and edited).

6. **Transition:** The design process concludes with a *transition* to migrate or release the mapping product to the user community. Too often, user

participation is included only at the transition stage through summative evaluation, which can lead to *feature creep*, or user requests for additional maps or mapping functionality. It also may lead to *feature loops*, or user requests that require redesigning existing maps or mapping functionality. The transition also should include a review of the original agreement between stakeholders and users to ensure that planned participation and accountability were achieved and that ownership and control are returned to the previously identified parties.

Table 7-1 provides a summary of this user-centered design process, alternative terminology used in related disciplines, and related milestones. This six-stage process outlines the basic structure for user-centered design, but every project requires a customized process to maximize user engagement under the given cost and time constraints. We recommend introducing and explaining the reasoning behind the process with stakeholders and users at the beginning of the project and adapting the stages and milestones based on input and feedback to ensure user participation throughout. Sometimes, users may have significant concerns about the project, such as levels of participation, accountability, and control, and may not be comfortable proceeding with the design process. Users also may grow to mistrust the project team and goals if insufficient effort is put into forming a meaningful and

Table 7-1. Six-stage user-centered design process

Stage	Alternative terminology	Milestones
Needs assessment	Task analysis, work domain analysis	User personas, use case scenarios
Conceptual design	Functional design, requirements analysis	Requirements document, comparative analysis
Prototyping	Brainstorming, mock-ups, proof-of-concepts	Low-fidelity wireframes, high-fidelity wireframes, specification sheet
User evaluation	Interaction studies, usability studies	User utility usability loops, focus groups, cognitive walkthroughs, online surveys
Implementation	Design, development, revision	Alpha (v0.1), beta (v0.2)
Transition	Debugging, editing	Final release (v1.0)

After Robinson et al. 2005; updated in Roth, Ross, and MacEachren 2015.

productive relationship based on mutually agreed-upon interests and values. Following user-centered and inclusive design, in such cases it is better to discontinue a project rather than continue without user participation and support.

Recommendations: Tactile map design considerations derived from user studies

User-centered and inclusive design is first and foremost a practical process for ensuring diverse user participation throughout design, with the aim of delivering a product that meets actual user needs. However, there are several critical moments during design when designers and users can slow the process to identify transferable design insights that might inform other, similar projects (see Roth 2019 for a review). Such studies have led to a growing number of empirically derived design recommendations for tactile mapping.

We summarize these insights in five categories of design recommendations that should be discussed with users during the needs assessment (stage 1 above), established as requirements in the initial conceptual design (stage 2), sketched into alternative design options during prototyping (stage 3), assessed for efficiency, effectiveness, and user preference in the user evaluation (stage 4), and implemented and transitioned into the final design product (stages 5 and 6). Thus, these design insights offer advice for getting the user-centered design process started based on prior user studies. However, they should not be considered as concrete rules prescribing design in all contexts. Instead, the final design should be based more on actual feedback from users and other budgetary, resource, and technological constraints than design recommendations derived from user studies.

- **Map composition and layout:** Use simplified layouts to enhance legibility; ensure each map symbol is distinguishable from its surroundings; maintain sufficient distances between map features and other map elements (typically 1–3 mm); avoid overlapping symbols.
- **Scale and generalization:** Maintain the relative size and position of map features for large-scale maps (that is, keep map features to cartographic scale); select only essential, task-relevant content for the map; reduce cognitive load where possible; apply height differentiation of symbols; create a series of simple maps with limited content for illustration of complicated phenomena.

- **Symbolization:** Use standardized height, spacing, and size for braille-based symbols; include standardized and evaluated tactile symbols where appropriate; use a small set of symbols and differentiate them using as many variables as possible (including height differentiation); use unique, easily identifiable symbols for the most critical map features; use 3D, volumetric symbols to enhance depth perception.
- **Interaction:** Enable touch-based input; enable haptic feedback such as vibration; combine haptic and auditory feedback; provide real-time updates on interactive displays; enable map customization to personal preferences.
- **User-centered and inclusive design process:** Enable people with visual impairments to participate in the design process; enable people with visual impairments to design their own maps.

Table 7-2 summarizes user studies and standards resulting in these empirically derived design recommendations to support future research and design on tactile mapping.

Following inclusive design principles, however, employing user studies for research purposes carries additional concerns and risks for users that must be carefully managed by designers and researchers. Participants should always give informed consent, and researchers should ensure a collective understanding of the project's purpose and possible benefits. At the time of this writing, at least 130 countries have legal protections for participants in user studies for research purposes, often requiring oversight by an institutional review board or similar ethics board (see the *International Compilation of Human Research Standards 2024*). In many contexts, it is recommended to cosign a memorandum of understanding (MOU) with partners who are supporting participant recruitment to ensure collective understanding of research benefits to participants and ownership of resulting research and design products.

Table 7-2. Tactile map design recommendations derived from user studies

Design recommendations	User study or standard
Map composition and layout	
Use simplified layouts to enhance legibility.	Hänßgen et al. (2016), BANA and CBA (2022)
Ensure each map symbol is distinguishable from its surroundings.	Giudice et al. (2020)
Maintain sufficient distances between map features and other map elements (typically 1–3 mm).	BANA and CBA (2022), Wabiński et al. (2022)
Avoid overlapping symbols.	BANA and CBA (2022)
Scale and generalization	
Maintain the relative size and position of map features for large-scale maps.	Hänßgen et al. (2016), Ottink et al. (2022)
Select only essential, task-relevant content for the map.	Wabiński et al. (2021)
Reduce cognitive load where possible.	Hofmann et al. (2022)
Apply height differentiation of symbols.	Edman (1992), Wabiński et al. (2022)
Create a series of simple maps with limited content for illustrating complicated phenomena.	Edman (1992), Wabiński et al. (2022)
Symbolization	
Use standardized height, spacing, and size for braille-based symbols.	BANA and CBA (2022)
Include standardized and evaluated tactile symbols where appropriate.	BANA and CBA (2022), (2010)
Use a small set of symbols and differentiate them using as many variables as possible (including height differentiation).	Hänßgen et al. (2016), Heni et al. (2016), Engel and Weber (2022), Wabiński et al. (2022), see also chapters 4 and 5 of this book
Use unique, easily identifiable symbols for the most critical map features.	Lobben and Lawrence (2012)
Use 3D volumetric symbols to enhance depth perception.	Gual et al. (2015), Bleau et al. (2023)
Interaction	
Enable touch-based input.	Heni et al. (2016), Palani et al. (2021)
Enable haptic feedback, such as vibration.	Heni et al. (2016), Palani et al. (2020), Rosenkranz and Altinsoy (2020), Palani et al. (2021)

Combine haptic and auditory feedback.	Brock et al. (2012), O'Modhrain et al. (2015), Hänßgen et al. (2016), Mascetti et al. (2016), Gorlewicz et al. (2020), Palani et al. (2020)
Provide real-time updates on interactive displays.	O'Modhrain et al. (2015), Krainz et al. (2016), Hofmann et al. (2022)
Enable map customization to personal preferences.	Hofmann et al. (2022)
User-centered and inclusive design process	
Enable people with visual impairments to participate in the design process.	Kitchin (2002), Rowell and Ungar (2005), Hänßgen et al. (2016), Brulé et al. (2020), Heni et al. (2016), Mascetti et al. (2016), Cole and Robinson (2023)
Enable people with visual impairments to design their own maps.	Shi et al. (2020)

Ethics: "Nothing about us without us"

"Nothing about us without us" is an advocacy slogan of the disability rights movement and an organizing principle of *disability justice*. This expression seeks to empower people with perceptual, cognitive, physical, or emotional disabilities to participate in, require accountability for, and take control over decision-making, policymaking, and in this case, design processes that affect them and their access to equal opportunities. Although the phrase has a long history, it is acutely linked to social movements in the 1980s and 1990s that led to policy, technology, and design improvements for people with disabilities. These included standards for braille printing and signage as well as web-accessible screen reader navigation and interaction. Importantly, disability justice reframes the medical model that disability is a dysfunction in individual bodies to a social-relational model that disability is culturally negotiated through decisions—design and otherwise—that privilege some bodies over others. In cartography, instead of designers brainstorming ways to modify existing mapping products designed for abled bodies, disability justice as a design tenet requires that users with disabilities participate in all stages of design to fundamentally rethink mapping products from their lived experiences and specific needs.

Design justice draws on disability justice as well as antiracist, decolonial, feminist, Indigenous, Marxist, and queer theory to seek participation from, require accountability for, and hand over control to users at the margins who are most affected by the designs (see Costanza-Chock 2020). Design is political, and each design decision made throughout the process distributes benefits and harms disproportionately across the multiple axes of intersectional user differences (Collins 1990). The Design Justice Network, a community of designers and researchers committed to design justice, has established a set of 10 principles to, among other intentions, make the distribution of benefits and harms more explicit and equitable across intersectional user differences in any design project (list 7-1).

List 7-1. Design Justice Network principles

1. We use design to sustain, heal, and empower our communities, as well as to seek liberation from exploitative and oppressive systems.
2. We center the voices of those who are directly impacted by the outcomes of the design process.
3. We prioritize design's impact on the community over the intentions of the designer.
4. We view change as emergent from an accountable, accessible, and collaborative process, rather than as a point at the end of a process.
5. We see the role of the designer as a facilitator rather than an expert.
6. We believe that everyone is an expert based on their own lived experience and that we all have unique and brilliant contributions to bring to a design process.
7. We share design knowledge and tools with our communities.
8. We work toward sustainable, community-led and -controlled outcomes.
9. We work toward nonexploitative solutions that reconnect us to the earth and to each other.
10. Before seeking new design solutions, we look for what is already working at the community level. We honor and uplift traditional, Indigenous, and local knowledge and practices.

The Design Justice Network (https://designjustice.org/) developed 10 principles to guide the design process, with the objective of promoting participation by users at the margin, requiring accountability by the design team, and empowering users to take control and ownership over the design process and products.

After Costanza-Chock 2020.

In and of itself, user-centered and inclusive design is not enough to ensure that the resulting designs are fair and just. For instance, many design innovations develop organically by users themselves through communities of practice. However, designers who mine these users for ideas through user-centered and inclusive design are typically rewarded with credit. It is not lost on us that we—three scholars who have corrected vision or color vision deficiency but, in the context of haptic mapping, are visually abled—will receive credit as authors of this chapter although the writing is based on the struggle, activism, and contributions of people with visual impairments. Even the most well-intentioned user-centered and inclusive design process can quickly become extractive of the very users the design is intended to support. Following design justice, each stage in the user-centered design

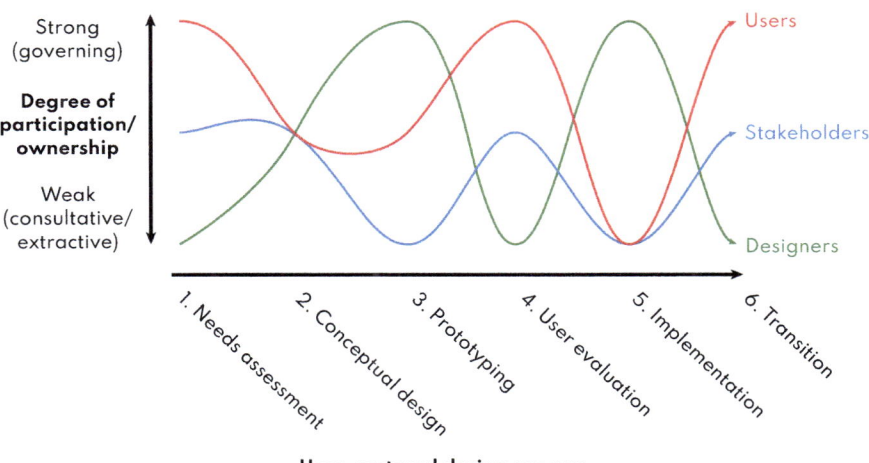

Figure 7-4. Monitoring ethically just participation during user-centered design. Participation in each stage of design can vary from strong, resulting in a governing or controlling role, to weak, resulting in a consultative or extractive role. The degree of participation for each designer, stakeholder, and user can then be traced as a line participating in the project. The figure shows hypothetical traces for users (red), other stakeholders (blue), and designers (green). These traces are provided for illustrative purposes and do not necessarily prescribe best practices. However, formalizing expectations about the degree of participation for each individual at the start before the needs assessment helps provide transparency and build trust among project contributors, as well as offering a concrete way to reflect if expectations are met throughout and after the project. Figure adapted from Costanza-Chock 2020; symbolization adapted from Kelly 2020.

process outlined above should be evaluated based on who is governing the process versus who is extracting value (figure 7-4).

Participation, accountability, and control look different depending on the goals of the project and the unique needs of the user community involved (Costanza-Chock 2020). It is important that users, designers, and stakeholders meet during the initial needs assessment phase. To make participation in the needs assessment accessible and effective, it helps to conduct these meetings in spaces familiar to the community.

Accountability can take the form of formal agreements, such as contracts and MOUs, signed by users at the outset to set clear expectations for everyone involved. Sharing requirements, wireframes, and other planning materials as editable, collaborative documents gives users the chance to contribute edits, circulate drafts for feedback, or adapt alternative designs for their purposes. For longer-term cooperations, creating advisory boards or councils that include user representatives facilitates ethical ownership over the final products.

Finally, precautions can be taken to reduce extractive elements of the user-centered design process. First, users should be compensated for their time and energy. In our experience, this means offering at least the local living wage for nonprofit or research projects and matching contractor rates for for-profit projects, plus covering travel expenses. Second, designers should seek ways to share credit with participating users. Recognition can include providing named attribution on the map or product, offering coauthorship for research papers, or sharing profits from commercialized products. Finally, designers should seek ways to elevate the visibility of users at the margins and their communities of practice, including speaking out against structural processes that push people to the margins and advocating for a more equitable distribution of benefits and a reduction of harms to all.

Further reading

On user-centered and inclusive design (outside cartography)
Braille Authority of North America and Canadian Braille Authority. 2010. *Guidelines and Standards for Tactile Graphics.*
Collins, Patricia Hill. 1990. *Black Feminist Thought, Knowledge, Consciousness, and the Politics of Empowerment.* Routledge.

Costanza-Chock, Sasha. 2020. *Design Justice: Community-Led Practices to Build the Worlds We Need.* MIT Press.

D'Ignazio, Catherine, and Lauren F. Klein. 2020. *Data Feminism.* MIT Press.

Microsoft Corporation. 2019. "Microsoft Inclusive Design." Accessed January 15, 2025. https://inclusive.microsoft.design/.

Newman, Damien. 2009. "That Squiggle of the Design Process." https://revisionlab.wordpress.com/that-squiggle-of-the-design-process/.

Nielsen, Jakob. 1992. "The Usability Engineering Life Cycle." *Computer* 25 (3): 12–22.

Norman, Donald A. 1988. *The Design of Everyday Things.* Basic Books.

Office for Human Research Protections. 2024. *International Compilation of Human Research Standards 2024 Edition.* Office for Human Research Protections (OHRP), Office of the Assistant Secretary for Health (OASH), US Department of Health and Human Services (HHS).

Rogers, E. M. 1962. *Diffusion of Innovations.* Free Press.

United Nations General Assembly. 1949. *Universal Declaration of Human Rights*, vol. 3381. Department of State, United States of America.

On user-centered and inclusive design (within cartography)

Bleau, Maxime, Camille van Acker, Natalina Martiniello, Joseph Paul Nemargut, and Maurice Ptito. 2023. "Cognitive Map Formation in the Blind Is Enhanced by Three-Dimensional Tactile Information." *Scientific Reports* 13 (1): 9736.

Brock, Anke, Philippe Truillet, Bernard Oriola, Delphine Picard, and Christophe Jouffrais. 2012. "Design and User Satisfaction of Interactive Maps for Visually Impaired People." *Proceedings*, Part II 13, Computers Helping People with Special Needs, 13th International Conference, ICCHP 2012, Linz, Austria, July 11–13, 2012.

Cole, Harrison, and Anthony Robinson. 2023. "Thematic Tactile Maps for Accessible Flood Mitigation Planning: Design and Evaluation." *Cartography and Geographic Information Science* 50 (6): 574–92.

Kelly, Meghan. 2020. "Feminist Mapping: Content, Form, and Process." PhD dissertation, Geography, The University of Wisconsin-Madison.

Robinson, Anthony C., Jin Chen, Eugene J. Lengerich, Hans G. Meyer, and Alan M. MacEachren. 2005. "Combining Usability Techniques to Design Geovisualization Tools for Epidemiology." *Cartography and Geographic Information Science* 32 (4): 243–55.

Roth, Robert E. 2019. "How Do User-Centered Design Studies Contribute to Cartography?" *Geografie* 124 (2): 133–61.

Roth, Robert E., Kevin S. Ross, and Alan M. MacEachren. 2015. "User-Centered Design for Interactive Maps: A Case Study in Crime Analysis." *ISPRS International Journal of Geo-Information* 4 (1): 262–301.

On user studies on haptic maps

Edman, Polly K. 1992. *Tactile Graphics*. American Foundation for the Blind.

Engel, Christin, and Gerhard Weber. 2022. "Expert Study: Design and Use of Textures for Tactile Indoor Maps with Varying Elevation Levels." International Conference on Computers Helping People with Special Needs.

Giudice, Nicholas A., Benjamin A. Guenther, Nicholas A, Jensen, and Kaitlyn N. Haase. 2020. "Cognitive Mapping Without Vision: Comparing Wayfinding Performance After Learning from Digital Touchscreen-Based Multimodal Maps vs. Embossed Tactile Overlays." *Frontiers in Human Neuroscience* 14:87.

Gorlewicz, Jenna L., Jennifer L. Tennison, P. Merlin Uesbeck, Margaret E. Richard, Hari P. Palani, Andreas Stefik, et al. 2020. "Design Guidelines and Recommendations for Multimodal, Touchscreen-Based Graphics." *ACM Transactions on Accessible Computing (TACCESS)* 13 (3): 1–30.

Gual, Jaume, Marina Puyuelo, and Joaquim Lloveras. 2015. "The Effect of Volumetric (3D) Tactile Symbols Within Inclusive Tactile Maps." *Applied Ergonomics* 48:1–10.

Hänßgen, Daniel, Nils Waldt, and Gerhard Weber. 2016. "Empirical Study on Quality and Effectiveness of Tactile Maps Using HaptOSM System." *Proceedings*, Part II 15, Computers Helping People with Special Needs, 15th International Conference, ICCHP 2016, Linz, Austria, July 13–15, 2016.

Heni, Saber, Wajih Abdallah, Dominique Archambault, Gérard Uzan, and Mohamed Salim Bouhlel. 2016. "An Empirical Evaluation of MoonTouch: A Soft Keyboard for Visually Impaired People." *Proceedings*, Part II 15, Computers Helping People with Special Needs, 15th International Conference, ICCHP 2016, Linz, Austria, July 13–15, 2016.

Hofmann, Megan, Kelly Mack, Jessica Birchfield, Jerry Cao, Autumn G. Hughes, Shriya Kurpad, et al. 2022. "Maptimizer: Using Optimization to Tailor Tactile Maps to Users' Needs." *Proceedings of the 2022 CHI Conference on Human Factors in Computing Systems*.

Kitchin, Rob. 2002. "Participatory Mapping of Disabled Access." *Cartographic Perspectives* 41:44–54.

Krainz, Elmar, Johannes Feiner, and Martin Fruhmann. 2016. "Accelerated Development for Accessible Apps–Model Driven Development of Transportation Apps for Visually Impaired People." Human-Centered and Error-Resilient Systems Development: *Proceedings* 8, IFIP WG 13.2/13.5 Joint Working Conference, Sixth International Conference on Human-Centered Software Engineering, HCSE 2016, and Eighth International Conference on Human Error, Safety, and System Development, HESSD 2016, Stockholm, Sweden, August 29–31, 2016.

Lobben, Amy, and Megan Lawrence. 2012. "The Use of Environmental Features on Tactile Maps by Navigators Who Are Blind." *The Professional Geographer* 64 (1): 95–108.

Mascetti, Sergio, Dragan Ahmetovic, Andrea Gerino, Cristian Bernareggi, Mario Busso, and Alessandro Rizzi. 2016. "Supporting Pedestrians with Visual Impairment During Road Crossing: A Mobile Application for Traffic Lights Detection." *Proceedings*, Part II 15, Computers Helping People with Special Needs: 15th International Conference, ICCHP 2016, Linz, Austria, July 13–15, 2016.

O'Modhrain, Sile, Nicholas A. Giudice, John A. Gardner, and Gordon E. Legge. 2015. "Designing Media for Visually-Impaired Users of Refreshable Touch Displays: Possibilities and Pitfalls." *IEEE Transactions on Haptics* 8 (3): 248–57.

Ottink, Loes, Bram Van Raalte, Christian F. Doeller, Thea M. Van der Geest, and Richard J. A. Van Wezel. 2022. "Cognitive Map Formation Through Tactile Map Navigation in Visually Impaired and Sighted Persons." *Scientific Reports* 12 (1): 11567.

Palani, Hari P., Paul D. S. Fink, and Nicholas A. Giudice. 2020. Design Guidelines for Schematizing and Rendering Haptically Perceivable Graphical Elements on Touchscreen Devices." *International Journal of Human–Computer Interaction* 36 (15): 1393–1414.

Palani, Hari Prasath, Paul D. S. Fink, and Nicholas Giudice. 2021. "Comparing Map Learning Between Touchscreen-Based Visual and Haptic Displays: A Behavioral Evaluation with Blind and Sighted Users." *Multimodal Technologies and Interaction* 6 (1): 1.

Rosenkranz, Robert, and M. Ercan Altinsoy. 2020. "Mapping the Sensory-Perceptual Space of Vibration for User-Centered Intuitive Tactile Design." *IEEE Transactions on Haptics* 14 (1): 95–108.

Rowell, Jonathan, and Simon Ungar. 2005. "Feeling Our Way: Tactile Map User Requirements—a Survey." International Cartographic Conference, La Coruna.

Shi, Lei, Yuhang Zhao, Ricardo Gonzalez Penuela, Elizabeth Kupferstein, and Shiri Azenkot. 2020. "Molder: An Accessible Design Tool for Tactile Maps." *Proceedings of the 2020 CHI Conference on Human Factors in Computing Systems.*

Wabiński, Jakub, Albina Mościcka, and Marta Kuźma. 2021. "The Information Value of Tactile Maps: A Comparison of Maps Printed with the Use of Different Techniques." *The Cartographic Journal* 58 (2): 123–34.

Wabiński, Jakub, Emilia Śmiechowska-Petrovskij, and Albina Mościcka. 2022. "Applying Height Differentiation of Tactile Symbols to Reduce the Minimum Horizontal Distances Between Them on Tactile Maps." *PLoS One* 17 (2): e0264564.

About the authors

Robert Roth is a professor in the University of Wisconsin–Madison (USA) Department of Geography and the director of the University of Wisconsin Cartography Lab (Cart Lab). Roth received a PhD in geography from the Pennsylvania State University in 2011. Roth is the current chair of the International Cartographic Association Commission on the User Experience. Roth researches user interface/user experience (UI/UX) design, mobile map design, and visual storytelling, among other topics. In the Cart Lab, Roth supervises undergraduate and graduate student internships on a range of print and digital client projects, many of which require compliance with web accessibility standards. In his free time, Roth enjoys walking his dogs, kayaking, and traveling.

Robert Roth.

Merve Keskin is a postdoctoral research associate in the Department of Geoinformatics (Z_GIS) at the University of Salzburg (PLUS), Austria. She is a geomatics engineer and cartographer, holding doctoral degrees from Ghent University (UGent) and Istanbul Technical University (ITU), obtained through a joint PhD program. She currently serves as vice chair of the International Cartographic Association Commission on User Experience. Her expertise includes spatial cognition and behavior, human sensing, the integration of HCI/UX/UI principles in map use and design, mixed-methods multimodal user experiments (eye tracking, EEG, EDA), and Explainable GeoAI. She supervises undergraduate and graduate students in these areas and strives to integrate new technologies and methods from various disciplines (psychology, computer science, urban planning) into geoinformatics through innovation and open science. She is the author of more than 35 peer-reviewed publications (see ORCID and Google Scholar). She enjoys making rap music and drawing in her free time. For more information, go to drmervekeskin.com.

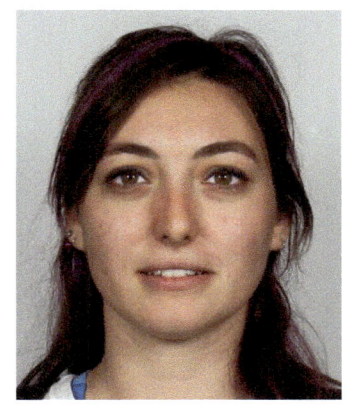
Merve Keskin.

Zdeněk Stachoň is an associate professor in the Department of Geography at Masaryk University (Czech Republic) and the head of the Laboratory for Virtual Geographic Environments. He currently serves as vice chair of the International Cartographic Association Commission on User Experience. Stachoň's research focuses on UI and UX design in cartography, with an emphasis on multivariate mapping, virtual geographic environments, and related topics. At the VGE Lab, Stachoň supervises undergraduate, graduate, and postgraduate students in a variety of research areas. In his free time, he enjoys cross-country running, mountain hiking, and traveling.

Zdeněk Stachoň.

Chapter 8

Learning geography when you're blind

Carla Cristina Reinaldo Gimenes de Sena and Waldirene Ribeiro do Carmo

Back in 1988, Carla and I began our studies in geography at USP—the University of São Paulo, Brazil. We were classmates but nothing more. Our interactions were limited to a "Good afternoon" in the hallway or an exchange of notes before a particularly daunting exam.

Fast-forward to 1995, when I started working as the technician in charge of LEMADI—the Laboratory for Geography Education and Teaching Materials at USP, renowned for its pioneering research in tactile cartography. Until then, I had hardly known that tactile cartography even existed! My first challenge was to learn to navigate something that felt so new and complex. And, fortunately, it was there that I met Carla again. She had interned at LEMADI during her undergraduate studies and was already well versed in the subject.

In 1996, Carla plunged headlong into her master's program, researching adaptations in environmental studies for people with visual disabilities. It was at this point that we truly became close. Through conversations about tactile maps and ideas for new adaptations, we discovered not only a shared passion for tactile cartography but also a deep and lasting friendship.

Our partnership blossomed: She went on to pursue her PhD, while I, already fascinated by the subject, worked up the courage to begin a master's degree, eventually followed by a PhD. Together, we worked on projects with researchers from Chile, Argentina, and Peru; delivered workshops, courses, and lectures; evaluated materials with students; and participated in various events. With each activity, we became increasingly aware of the profound impact of tactile cartography on geography education and citizenship.

Now, as we share this story, we do so in the hope that it inspires others to dream of a more accessible world, where learning is a right for all.

—Waldirene

Introduction

Teaching children and young people in schools to grasp the core concepts of geography, including place, landscape, region, and territory, is as fundamental as teaching them subjects such as mathematics or history. Yet geography textbooks often use expressions such as "observe the landscape" or "describe what you see from the classroom window," with a predominantly visual focus. However, over time, teaching methods have evolved, and teaching geography in schools today should go beyond simply describing landscapes or memorizing names and locations. It must foster a critical understanding of how natural and social elements interact to shape the world we live in.

To fully understand geography, students need to consider both the physical aspects, such as climate, relief, soil, and vegetation, and the human aspects, such as economic activities, population distribution, cultural issues, and geopolitical challenges. The subject thus contributes to the development of an integrated and critical perspective, enabling students to comprehend the processes influencing—for instance, environmental issues and contemporary global challenges. In a world shaped by inequality and rapid change, it is crucial to move beyond mere observation of space and delve into the processes that shape it.

Among the various languages geography employs to convey its concepts and interpretations, cartographic language holds particular importance. It permeates the learning process for all geography content, enabling the synthesis of information, the expression of knowledge, and the study of how space is produced, organized, and distributed.[1]

Understanding a map involves mastering the concepts inherent in its creation, such as the scale that establishes proportions between reality and its representation, the geographic coordinates used for locating places, and the symbology applied to represent each mapped element. Teaching students how to map, and how to read maps, requires careful attention from educators and must include consideration of all learners, including those with disabilities.

Most graphic representations available are in printed or digital formats, relying exclusively on the sense of sight, which restricts learning to a single sensory channel. This predominantly visual approach can exclude multisensory learning and, consequently, visually impaired students. It is therefore essential to create images and graphic representations adapted to tactile formats. Tactile cartography plays a pivotal role in this effort, providing materials that support an inclusive approach. In this way, all students, regardless of their visual abilities, can explore and understand geographic content in an integrated and accessible manner.

Tactile cartography as a science is very young and its main concern is to transform visual maps into a product that can be used by all people, including the visually impaired. However, relief maps have been produced for a long time, and some reports of tactile maps from the 19th century illustrate the importance of looking for alternatives in the teaching of geography (see chapter 1, "From Visual to Tactile: Societal Attitudes and Accessible Information").

From an inclusive perspective, in the broadest sense of the term, the use of tactile maps in geography teaching can ensure that students with visual impairments gain access to curriculum content, enabling meaningful learning and the practical application of knowledge in their daily lives. To accomplish this, it is essential to adapt visual materials into tactile and audio-tactile formats to communicate information effectively while also teaching students how to read and interpret these maps.

In this chapter, we emphasize a methodical approach to learning cartographic concepts and the role of those concepts in teaching geography to people with visual impairments. Additionally, we highlight the importance of training teachers from an inclusive perspective. There is little point in having excellent tactile maps if teachers simply put them away because they do not feel confident using them.

Teaching map reading to students with visual impairments

Throughout human history, people with disabilities have faced various forms of exclusion and segregation. It was only in the second half of the 20th century that significant initiatives toward inclusion began to be discussed in various parts of the world. The 1990s marked a turning point, with many countries committing to develop tools that supported education for people with disabilities within the framework of the inclusion paradigm. This approach emphasizes society's responsibility to create equal opportunities for all individuals.

Students with disabilities, who had previously been unable to attend school or did so in a segregated manner, began to join mainstream schools. This shift demanded changes not only in the architecture of school buildings but also in teaching methodologies, content selection, and, most importantly, teaching resources.

For students with visual impairments, the use of resources adapted for tactile perception is essential to ensure that learning the concepts and themes of geography does not rely solely on oral descriptions. Geography is, of course, an essentially visual science that employs graphic representations, particularly maps, to convey its analyses. Yet, as a complex representation governed by rules and requiring a certain level of abstraction to comprehend, maps can shift from being valuable allies in geography teaching to becoming significant barriers to effective learning if they are not treated as a subject that needs to be explicitly taught.

By closely examining the cartographic process of creating a map, we can identify various concepts embedded in this method of conveying spatial information. A map is developed from a perspective that is

- vertical (think of a house: it's shown on the map from above, but in everyday life we're used to seeing it from the front or from the side)
- generalized (because the map is a reduction, we can't show all objects, and we must select and simplify, which is even more important in tactile maps)
- scaled (which helps us understand the reduction used in the map and thus establish the relationship with reality)
- georeferenced (all the places represented on the map have their location based on coordinates, such as our address, which crosses the street name with the house number)

Additionally, a map uses symbols to represent a specific location and, beyond location, can encompass a wide array of themes.

Considering the abstraction required to understand a cartographic representation, it is important to develop and implement a program to introduce the use of maps in geography education. Methodical practices aimed at working with the basic concepts of maps should be carefully planned to promote the development of cartographic concepts. This is no different for students with visual impairments. It is not enough to adapt maps into a tactile version if there is no methodical approach focused on teaching cartographic concepts, ensuring that the adapted maps are effectively used as tools for teaching geography.

To achieve this goal, we recommend working on the following notions through a variety of activities tailored to the age group and level of schooling: different points of view (oblique and vertical vision); three-dimensional and two-dimensional images; symbology (point, line, and area); the construction of legends, proportion, and scale; laterality for understanding geographic orientation and location (geographic coordinates).

Without these introductory activities, most basic education students are unable to understand the information represented on maps. For students with visual impairments, the situation is even more critical, particularly because teaching materials adapted to the specific needs of people with visual impairments are generally unavailable in schools, especially in poorer countries. In this context, the introduction to tactile graphic language is a crucial step in developing cartographic understanding for students with visual impairments. The introductory activities demonstrate to visually impaired students how different textures and reliefs can be used to represent spatial information.[2]

Proposals for activities to introduce the use of maps

To represent in two dimensions what is observed in three dimensions, it is crucial to understand the vertical view, which accounts for length, width, and height from a top-down perspective.

For students with visual impairments, we propose the creation of relief figures with different textures that represent everyday objects from various points of view (horizontal and vertical). This should start with simple objects that can be touched, such as glasses, bottles, cutlery, toothbrushes, or even a smartphone, represented in their real size. Gradually, this can progress to larger objects that require proportional reduction, such as a house, boat, or soccer field.

The use of figures with different textures aids in tactile training, fostering future understanding of the elements represented on maps and their corresponding definitions in the legend. Additionally, it introduces the concept of proportion, which is essential for the effective use of scale.

Scale

Scale, which represents the relationship between the measurements on a map and the actual measurements of the object or space being mapped, requires children to grasp the notion of proportion. It is important for them to understand that an object or area can be represented in different sizes and that the choice of map scale depends on the level of detail intended to be conveyed, based on specific objectives. This understanding helps children recognize how scale influences the way space is portrayed and interpreted.

For tactile cartography, the concept of scale is essential because people with visual impairments read the map through a scanning process, perceiving each part as they touch it. Their understanding of what is represented is gradual, requiring a higher level of abstraction than that of a sighted person who can see the entire map at once (see chapter 3, "Understanding Through Touch").

By understanding that this tactile map is proportionally reduced in relation to reality and grasping the extent of this reduction, people with visual impairments can more efficiently establish comparisons of distances and different levels of generalization. This is particularly important since most maps used in geography education are small scale, covering large areas to present various themes. When you zoom out (small scale), you see more of the scene but with less detail. When you zoom in (large scale), you see less of the scene but with more detail.

Once again, everyday objects serve as the starting point for working with the concept of scale. For example, a car can have its measurements compared with toys of different sizes, just as human or animal dolls can be compared with the measurements of an adult. These objects are represented in tactile models of different sizes, with the scale used clearly identified. Tactile *maquettes* complement these activities by representing larger objects that require significant reduction and generalization of shapes and details, such as a house, park, river, or mountain. In research conducted over the last 30 years in Latin America, the use of the graphic scale has been emphasized, replacing the numeric scale, which represents a considerable simplification. The graphic scale is considered more accessible because it is less abstract and allows for an immediate comparison of the reduction applied.

Orientation

Everyone, to some extent, has had to navigate somewhere. Identifying landmarks and using the geographic coordinate system enables us to orient ourselves, in relation to both the space around us and other places. For people with visual impairments, practicing spatial orientation is crucial for greater independence in their daily activities. Although the reference points in this case are not visual, they remain essential for safe movement.

People with visual impairments can use characteristic sounds of certain places, floor textures, smells, or even differences in room temperature as references. The combination of these sensory elements helps create a "mental map," which enables them to navigate more independently. Additionally, orientation and mobility training activities, as well as the use of tactile maps and models, are vital resources for people with visual impairments to develop and improve their ability to orient themselves and interact with the space around them. For more information about orientation and mobility, see chapter 9 ("Training in Orientation and Mobility").

Coordinates

Learning the coordinate system is essential for understanding geographic phenomena, establishing spatial relationships, and solving problems related to location, distribution, and interaction between elements in space. Beyond geography lessons, mastering the use of geographic coordinates will facilitate the practical application of knowledge in everyday situations and in studying the organization of space, enhancing spatial reasoning.

To include people with visual impairments in learning geographic coordinates, we need to understand that these coordinates are based on imaginary lines drawn on the globe. This requires the use of tactile globes, which allow students to grasp parallels and meridians, as well as the degrees of latitude and longitude necessary for more precise location identification in the representation we make of the globe on maps. In addition, playful strategies, such as educational games, can make the learning process more accessible and dynamic.

One example is the adaptation of the classic Battleship game into a "geographic battle," where a network of intersecting vertical and horizontal lines is created using textures. Initially, the board represents a simple coordinate network, using letters and numbers to identify positions. As learning progresses, the game can be expanded to include a more complex grid, based on the Earth's hemispheres and specific latitude and longitude coordinates. The "pieces" of the ships come in different sizes and textures and can represent other objects or even elements of the landscape, such as an island or a mountain, giving the game its geographic character.

Symbology of the map

Graphic representation consists of a system of signs organized for us to understand and communicate information. This process of graphic communication is directly influenced by the characteristics of the phenomenon being represented and by the available data. Each cartographic symbol serves the purpose of representing objects, phenomena, or features present in geographic space, such as rivers, mountains, cities, and borders.

To be effective, symbols must be clear, understandable, and consistent with cartographic language. The choice of symbols reflects the purpose of a map through a specific composition of signs. Graphic semiology[3] studies this system of signs and provides guidance on evaluating the advantages and limitations of visual variables used for representation. It also defines the rules of cartographic language based on visual perception. Graphic semiology enables the logical and aesthetic expression of any phenomenon on the Earth's surface. However, because its components are predominantly visual, these characteristics need to be adapted for tactile perception.

In tactile adaptation, distinguishing among objects and phenomena—typically achieved through variations in points, lines, and areas defined by shape, size, and color—incorporates additional variables, such as height and texture.[4] These adaptations define shapes and sizes three-dimensionally while replacing color as a means of differentiation. The legend, essential for communicating information on printed or digital maps, becomes even more strategic in tactile maps. It not only helps students understand the content of the map but also serves as a tool for tactile training.

It is important to highlight that tactile maps, when combined with visual information adapted for individuals with low vision, become resources accessible to all students (figure 8-1). This approach fosters inclusive learning and encourages interaction between students with and without disabilities.

The "city game"

To systematize cartographic concepts, we present one of the proposed activities,[6] which has been used in courses and workshops for teachers across various areas of basic education in Brazil and some Latin American countries. When applied in geography classes, these structural concepts broaden the range of information available to students with visual impairments about spatial phenomena and representations.[7]

The "city game" (figure 8-2) serves as a synthesis of other activities related to the introduction of tactile maps, having been evaluated by people with visual impairments and primary school teachers. The game consists of a set of nine 5 × 5 cm pieces, each with a distinct shape and texture, representing elements in the center of a small town (town hall, school, theater, police station, ice-cream parlor, hospital, church, library, and house). It also includes a larger 15 × 15 cm piece symbolizing the town square. All pieces are magnetized to be placed on a metal board. The game features a tactile compass rose, which can be supplemented with a braille compass, as well as a legend to identify each piece.

Learning geography when you're blind   163

Figure 8-1. Evaluation of models that associate tactile information with colors for use in inclusive classrooms.[5]

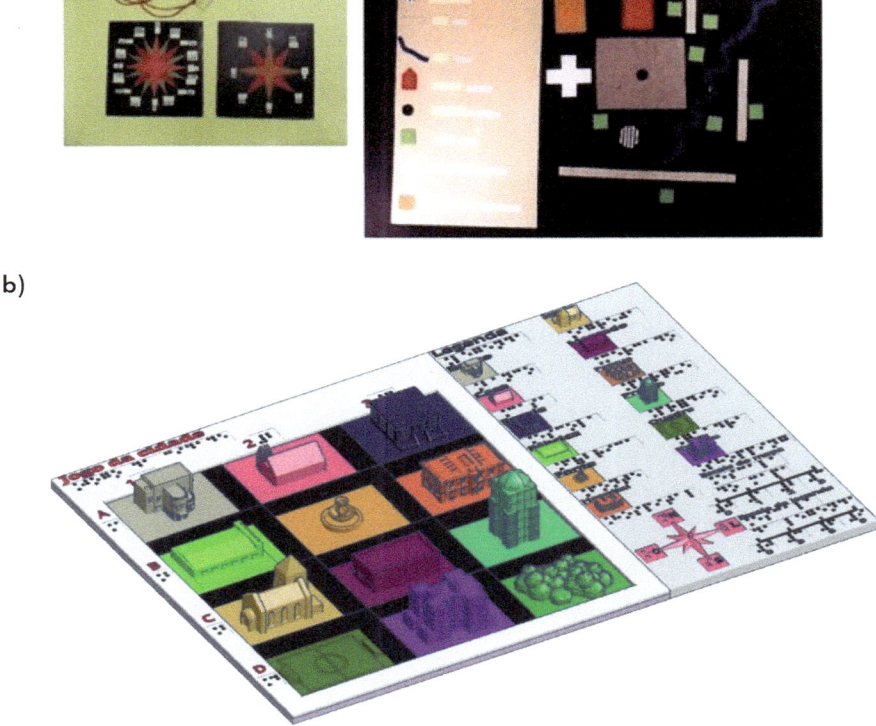

Figure 8-2. (a) A handmade model of the city game and (b) a study of how the game could be made using 3D printing.

With the square placed at the center of the board, the student is challenged to position each piece according to an orientation, following the cardinal and intercardinal points. The objective is to prepare the student for the use of visual and tactile graphic representations and to develop the student's ability to differentiate textures, colors, shapes, and sizes. The activity also promotes the learning of orientation and location concepts and introduces the use of the legend. This, in turn, develops students' ability to relate generalized, reduced, and schematic representations of the map to their corresponding real-world counterparts.

Introduction to the use of tactile maps helps students read and interpret these maps more effectively, helping them, in the words of Barbara Gomes Flaire Jordão, to understand part of the reality in which they live and thus be able to act on it:

> When they feel safe, in possession of a theoretical-conceptual framework based on the use of representations that stimulate their spatial reading, they understand spatial elements, develop skills, look for new attitudes, and gain autonomy to act and think about space. As a result, students are ready to take their discussions beyond common sense and meet the science of geography.[8]

Tactile representations can be designed and organized in a rational manner, but their effectiveness relies on being interpreted correctly and applied in the appropriate context. This process does not conclude with the cartographer's work; it must be complemented and enriched through collaboration with educators.

Illustrations of successful training programs

Geography teaching occurs within a broad context that encompasses its theoretical and methodological foundations, the curricular structure of official documents, the reality of the school environment, the daily lives of students, and teacher training. In Brazilian schools, teachers have had successful experiences creating tactile maps for students with visual impairments in the final years of elementary school, in classrooms with sighted students who also use these maps. These experiences are structured around working with map elements and their importance in conveying the intended information. Activities such as the city game encourage discussions among the students about alternatives that would allow people with visual impairments to also use maps and participate in geography lessons.

Craft techniques (figure 8-3), particularly collage, are used to create tactile maps on a variety of themes, revisiting cartographic concepts, which are then applied to new maps. By making tactile maps, children and young people learn cartography in an efficient and meaningful way, while also helping to address the lack of inclusive resources for their peers.

These activities have spread to schools in Brazil and in other countries in Latin America because of the numerous courses and workshops we have conducted over the past 20 years, which have encouraged teachers to work with tactile maps created

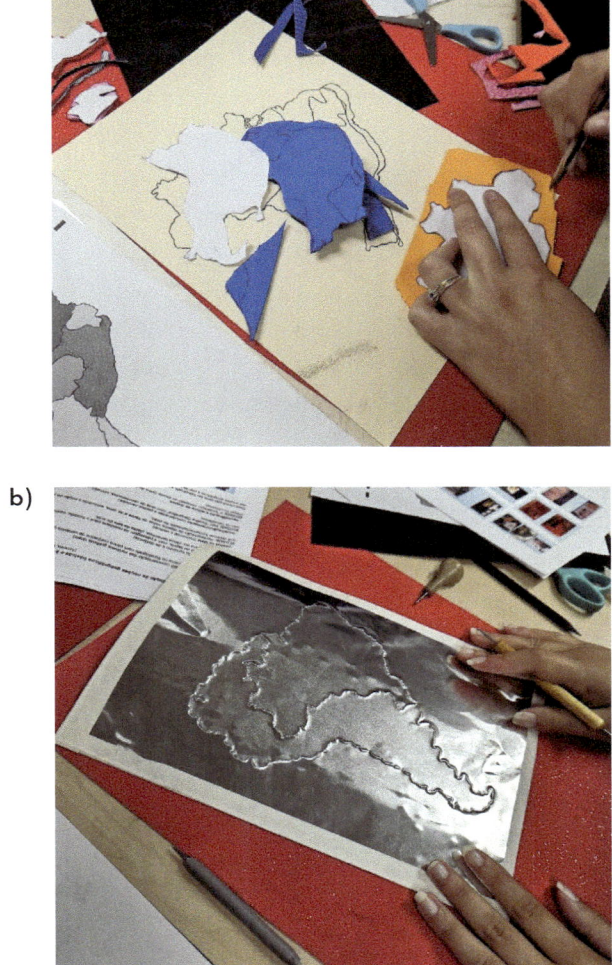

Figure 8-3. Workshops are the best way to spread knowledge about tactile cartography and geography education. Participants in the workshop Inclusive Cartography at the Brazilian Congress of Geographers, in São Paulo, Brazil, in July 2024, (a) do collage and (b) apply the shapes to new maps.

by their students. This approach has made lessons more dynamic and has facilitated the understanding of various cartographic concepts.

Reports from teachers during these courses and workshops indicate an improvement in the performance of the children and young people involved in the project, an increased interest in maps in general, a greater awareness of accessibility and inclusion issues, and a notable rise in the production of tactile maps in schools.

Conclusion

Teaching geography to people with visual impairments presents a challenge, but it also offers an opportunity to redefine cartography and its significance in school geography. Three-dimensional models encompass key concepts, such as location, proximity, association, adjacency, distance, direction, orientation, and other essential spatial properties of the real world. These models serve as a means for people with visual impairments to experience the geospatial domain.

Tactile maps and models not only contribute to geography teaching in general but also play an especially important role in conveying the highly abstract concepts of physical geography. Explaining the difference between plateaus and plains, for instance, or illustrating mountain types or climate dynamics becomes more meaningful with multisensory resources. This approach benefits all classrooms, regardless of whether a person with disabilities is present.

Tactile maps are powerful teaching tools for geography. They can be created using craft techniques and low-cost materials, as well as with technological resources such as software and 3D printers. However, what truly makes a difference in learning is not the maps themselves but the development of methods that use them in a meaningful and contextualized way.

When selecting tactile maps to teach geography, it is essential to consider both their specificities and the characteristics of the target audience. For this reason, tactile cartography should be incorporated into initial and ongoing teacher training, with a focus on working with each element of the map and its fundamental concepts.

School geography fosters critical skills, such as problem-solving, spatial thinking, and analytical reasoning, all of which contribute to the development of geographic reasoning. Making these skills accessible to individuals with disabilities broadens their scope, promoting inclusion and equality of opportunities. Achieving this accessibility requires the adaptation of materials, methods, and teaching practices, ensuring that all students can comprehend geographic concepts and engage meaningfully with the space around them.

Further reading

In this section, we have added texts that complement or expand on the themes presented in the chapter, which are subdivided into school cartography and tactile cartography. Much research has been carried out on tactile cartography and the use of maps in schools by children and young people in Brazil, which is recognized for its research in these areas—but most of the publications are in Portuguese. We believe that these are important contributions to the subject, and that's why we invite you to get to know these texts. Come on, choose your favorite translator and immerse yourself in the way Brazilian researchers think about tactile cartography. Who knows? Maybe you'll learn a few words in Portuguese and visit us in Brazil! Enjoy your reading! (*Boa leitura!*)

On school cartography

Almeida, Rosângela Doin de. *Do desenho ao mapa: Iniciação cartográfica na escola*, 4th ed. São Paulo: Contexto, 2003.

Almeida, Rosângela Doin de, and Elza Y. Passini. *O espaço geográfico: Ensino e representações*. São Paulo: Contexto, 2008.

Almeida, Rosângela Doin de, Miguel César Sanchez, and Adriano Picarelli. *Atividades Cartográficas*, vols. 1–4. São Paulo: Atual Editora, 1996.

Ferreira, Graça Maria Lemos, and Marcelo Martinelli. *A Caminho dos Mapas*, vols. 1-5. São Paulo: Editora Moderna, 2014.

Pissinati, M. C., and R. S. Archela. "Fundamentos da alfabetização cartográfica no ensino de Geografia." *Geografia (Londrina)* 16, no. 1 (2010): 169–95.

Reinaldo Gimenes de Sena, Carla Cristina, and Waldirene Ribeiro do Carmo. "School Cartography in Brazil and its Inclusive Perspective." *International Journal of Cartography* 6 (2020): 316–30.

Simielli, Maria Elena Ramos. *Primeiros mapas: Como entender e construer*, vols. 1–4. São Paulo: Ática, 1993.

On tactile cartography

Andrews, S. K. "Applications of a Cartographic Communication Model to Tactual Map Design." *The American Cartographer* 15, no. 2 (1988).

Carmo, Waldirene Ribeiro do. "Cartografia tátil escolar: experiências com a construção de materiais didáticos e com a formação continuada de professores." Dissertation (master's degree in geography), Department of Geography, Faculty of Philosophy, Languages and Literature, and Human Sciences. University of São Paulo, Brazil, 2009.

Castner, Henry W. "Tactual Maps and Graphics: Some Implications for Our Study of Visual Cartographic Communication." *Cartographica* 20 (1983).

Edman, Polly K. *Tactile Graphics*. American Foundation for the Blind, 1992.

Eriksson, Yvonne. *Tactile Pictures: Pictorial Representations for the Blind, 1784–1940*. Gothenburg: Acta Universitatis Gothoburgensis, 1998.

Eriksson, Yvonne. "What Is the History of Tactile Pictures?" In *Art Beyond Sight: A Resource Guide to Art, Creativity, and Visual Impairment*, by Elizabeth S. Axel and Nina S. Levent. American Foundation for the Blind, 2003.

García, Fernando, and Pedro Ruiz. "Mapas geográficos para personas ciegas y deficientes visuales." *Integración: Revista sobre discapacidad visual*, Espanha: ONCE, no. 57 (2010): 56–72. Accessed 15 Sep. 2024. https://dialnet.unirioja.es/servlet/articulo?codigo=5828913.

Golledge, Reginald G. "Learning Geography in the Absence of Sight." In *WorldMinds: Geographical Perspectives on 100 Problems*, by Donald G. Janelle and Barney Warf and edited by Kathy Hansen. Springer, 2004.

Tatham, A. F. "The Design of Tactile Maps: Theoretical and Practical Considerations." In *Proceedings of the 15th Conference Mapping the Nations* 1. ICA, Bournemouth, England, 1991.

Tatham, A. F. *Como confeccionar mapas y diagramas en relieve, Los Ciegos en el Mundo*. Madrid: Union Mundial de Ciegos, 1993.

Ungar, S. "Cognitive Mapping Without Visual Experience." In *Cognitive Mapping: Past, Present, and Future*, edited by R. Kitchin and S. Freundschuh. Routledge, 2000, 221–48.

Vasconcellos, Regina (Araujo Almeida). "A Cartografia Tátil e o Deficiente Visual: Uma avaliação das etapas de produção e uso do mapa." Doctoral thesis (Doctor of Geography), Department of Geography, Faculty of Philosophy, Languages and Literature, and Human Sciences. University of São Paulo, Brazil, 1993.

Wiedel, Joseph W. *Virkotype Process of Raised Printing*. Division for the Blind, Washington: Library of Congress, Circular 64-10 (with Alfred Korb), 1964, 256.

About the authors

Born in São Paulo, Brazil, Carla Cristina Reinaldo Gimenes de Sena has always been passionate about travel and an avid reader of adventure books. She holds a degree in geography and has worked with children and young people for 19 years in basic education. During her undergraduate studies, she became acquainted with the tactile cartography project, which later became the focus of her master's and doctoral research at the University of São Paulo (USP), Brazil. She is currently an associate professor at the Department of Geography at the Faculty of Science, Technology, and Education (FCTE) at São Paulo State University. Her research primarily focuses on teaching methodologies in geography and cartography, particularly inclusive cartography, and she supervises undergraduate and graduate students in these fields. Additionally, she holds the position of vice-director of FCTE at the university. Carla began participating in International Cartographic Association (ICA) conferences in 2005, served as the chair of the Maps and Children Commission from 2015 to 2023, and currently collaborates with the Inclusive Cartography working group.

Carla Cristina Reinaldo Gimenes de Sena.

Waldirene Ribeiro do Carmo is a geographer who graduated from USP, Brazil, where she also earned her master's and PhD in sciences (physical geography). She is the researcher in charge of the Laboratory of Geography Education and Teaching Materials at the Geography Department of the Faculty of Philosophy, Languages and Literature, and Human Sciences at USP. She is also a collaborating researcher at the Center for Tactile Cartography at the Universidad Tecnológica Metropolitana in Santiago, Chile. She served ICA's Commission on Maps and Graphics for Blind and Partially Sighted People as cochair from 2011 to 2019 and as chair from 2019 to 2023. Currently, she is a member of the ICA's Inclusive Cartography

Waldirene Ribeiro do Carmo.

working group. Her areas of expertise include teacher training in school and tactile cartography, special education, and inclusion.

Notes

1. Jordão, Barbara Gomes Flaire, "O pensamento espacial e o raciocínio geográfico em alunos com deficiência visual: O papel da Cartografia Tátil," Doctoral thesis (doctor of geography), Department of Geography, Faculty of Philosophy, Languages and Literature, and Human Sciences (University of São Paulo, Brazil, 2021).
2. Reinaldo Gimenes de Sena, Carla Cristina, and Waldirene Ribeiro do Carmo, "School Cartography in Brazil and Its Inclusive Perspective," *International Journal of Cartography* 6 (2020): 316–30.
3. Bertin, Jacques, *Sémiologie Graphique: Les Diagrammes, Réseaux, les Cartes*, 10th ed. (Paris: Monton & Gauthier Villars, 1973).
4. Vasconcellos, Regina (Araujo Almeida). "A Cartografia Tátil e o Deficiente Visual: Uma avaliação das etapas de produção e uso do mapa." Doctoral thesis (Doctor of Geography), Department of Geography, Faculty of Philosophy, Languages and Literature, and Human Sciences. University of São Paulo, Brazil, 1993.
5. Ribeiro, Diego Alves, "Produção de Recursos Didáticos Inclusivos para o Ensino de Geografia por meio da Impressão 3d." Dissertation, master's degree in geography (State University of Londrina, Paraná, Brazil, 2024).
6. Vasconcellos, doctoral thesis, 1993.
7. Jordão, doctoral thesis, 2021.
8. Jordão, doctoral thesis, 2021.

Chapter 9

Training in orientation and mobility

Petr Červenka

In the mid-1990s, we installed a model and orientation plan of the surroundings of the Prague School for the Blind. Our colleague, the school's orientation and mobility teacher, became impatient and discouraged, as the students started lamenting, "It's too complicated!...I don't know how to do this!...Just tell me which way to go!" We encouraged her to persevere.

Months later, the school contacted us with an unexpected request. They needed a map of the distant surroundings, as the older students had already mastered the immediate area. But some students now refused to leave their familiar territory, insisting, "I won't go somewhere I don't know." At that moment, we knew that we were on the right track, because the students clearly felt confident in their immediate surroundings. Our initial model had done its job. We delivered the requested map immediately, and the more distant surroundings of the school opened up to the students for exploration and orientation.

—Petr

Micro-orientation and macro-orientation

Navigating unknown spaces without sight is incredibly challenging, often leading to dependency and isolation for people with visual impairments. Institutions for the blind and partially sighted invest heavily in orientation and mobility (O&M) training because this empowers people with visual impairments with the necessary skills to gain independence and mitigate social isolation. It is an important part of the rehabilitation process.

Orientation involves gathering and using information from the surrounding environment to understand spatial relationships and plan movements.[1] Here it is important to distinguish two concepts: micro-orientation and macro-orientation. Micro-orientation refers to the process of understanding and navigating smaller, immediate spaces. It relies primarily on the sense of touch, specifically the tactile field. This can involve using one hand (mono-manual) or both hands (bimanual) to explore objects and surfaces—for example, feeling a textured sidewalk or a doorknob.

Macro-orientation, in contrast, involves navigating larger spaces that extend beyond the immediate reach of the individual's senses. This requires the use of distant senses, such as hearing and echolocation—for example, using sound cues to determine the distance and direction of a sound source or using a cane to detect obstacles. In essence, micro-orientation is about the immediate, whereas macro-orientation is about the distant. Both are essential to moving safely and independently in an environment. People with visual impairments are considered mobile when they can confidently and safely navigate spaces, using acquired movement techniques and sensory information.

The O&M teacher

The work of O&M teachers is highly demanding because it requires a comprehensive understanding of various scientific disciplines, including pedagogy, psychology, medicine, and technology, as well as the ability to apply this knowledge effectively in practice. This role necessitates not only mastering these foundational areas but also adopting a sensitive and individualized approach to each client. O&M teachers must be adept at working with people of different ages, professional backgrounds, intellectual levels, and value systems, while also respecting the specific etiology of each visual impairment.

According to a proposal from a European seminar on O&M education, the O&M trainer is a specialized teacher who develops the motor, sensory, and mental skills of people with visual impairments to help them achieve independent, safe, and efficient travel.[2] This includes evaluating, planning programs and lessons, and documenting training. The trainer ensures client safety, recognizes the importance of independent movement, helps individuals reach their full potential, and adapts instructions to meet each client's needs. Additionally, the trainer understands the medical aspects of visual impairments, translates medical terminology into practical instructions, demonstrates practical activities, explains mobility aids, and suggests environmental modifications. The preparation for trainers involves a minimum of 300 hours of direct teaching for postgraduate students over an extended period.

Basic areas of O&M education

Working with an O&M teacher, people with visual impairments require systematic training in orientation and mobility skills, defined as habits that enable people with visual impairments to achieve a high level of mobility.[3] These habits can be divided into mobility and developmental techniques. Among basic mobility techniques are the training of people with visual impairments to walk with the assistance of a sighted guide, adopt positions that ensure safety, and use their hands to follow trails along walls or other surfaces for guidance. Developmental techniques develop natural orientation abilities, such as minimizing the tendency to drift off course, estimating the distance and angles of objects, perceiving and navigating curves, recognizing and adapting to changes in terrain slope, using sound to navigate, developing an awareness of obstacles, and improving balance and stability.

Long-cane technique

A well-known O&M aid is the white (long) cane. According to Hoover (1950), a cane has two functions: to bumper and to probe.[4] Additionally, it informs others, such as passers-by and drivers, that the user is visually impaired, prompting them to be cautious and assist if needed. The long cane provides a level of safety and protection and makes it easier to locate orientation points (specific locations used for navigation) and tactile signs (physical markers that can be felt and used for guidance).

It is usually recommended that the cane's length be equal to the distance from the floor to the middle of the user's chest. When the cane is primarily used for daily activities, such as commuting and navigating crowded areas, a longer cane could become cumbersome and ineffective. Even with an optimal center of gravity, it could be difficult to manage, especially in public transport or crowded places. Users might also need to adjust the length of the cane frequently, making the extra length impractical. A cane properly selected for the user's height ensures safety, confidence, and adherence to physiological and aesthetic principles of movement.

Canes are usually provided in two variants. A supportive cane is for short-distance support and needs to be sturdy and possibly foldable, whereas a long cane is for achieving high mobility and thus should be light and easy to handle. In O&M training, the focus is on practicing the long-cane technique.

Training long-cane techniques

Learning how to master the long cane consists of several phases. The first phase focuses on teaching the basic techniques, which can be done in a larger room with sufficient clear space for safe movement. These techniques include the key elements of using a cane for O&M, such as a correct posture, holding techniques, and the coordination of cane movements with body movements. Practicing these techniques with both hands ensures better proficiency and adaptability.

The second stage focuses on practicing and solidifying these activities in the repertoire of people with visual impairments to build confidence and proficiency in various walking scenarios, from simple routes to more complex ones with obstacles and gradient changes. It also includes specific practice for stair navigation and walking in areas without clear orientation points. The goal is to ensure that these activities become a natural part of the individuals' everyday lives, enhancing their overall mobility and independence.

In the third phase, the focus is on practicing orientation activities that require the integration of individual analyses. The goal is to handle typical situations effectively. Training is conducted in urban environments, starting in calm areas and progressing to busier centers; it involves navigating simple and more difficult routes, both with and without obstacles, and finding specific targets. The progression from calm to busy areas helps build confidence and proficiency in real-world situations.

The last and most difficult stage of O&M training involves using all available information and techniques to navigate effectively. This includes using senses other than sight, such as hearing, touch, smell, and proprioception (awareness of body position). It also involves developing innate abilities to understand and navigate the environment, using aids such as long canes and tactile maps, and following verbal or written instructions. The goal is to teach people with visual impairments to gather and use information optimally, form an accurate understanding of their environment, and develop effective navigation solutions through practice.

Importance of O&M training

It is crucial to teach people with visual impairments to identify important phenomena for orientation and assess their significance. Sometimes there are many of these phenomena, and sometimes there are few, so it is important to teach people with visual impairments to identify and prioritize these phenomena. Incorporating the terms *orientation point* and *orientation sign* into O&M training helps with environmental descriptions and improves the confidence and effectiveness of people with visual impairments.

An orientation point is a distinct and easily recognizable feature that remains constant in its location and shape. It must be easily detectable while walking and provide new information, facilitating the orientation of people with visual impairments. Orientation points must be reliable in all seasons and weather conditions and must not be easily confused with one another, making isolated pillars, lampposts, or obstacles such as mailboxes unsuitable. Suitable points are those that can be found regardless of climate changes

Orientation signs are phenomena that help visually impaired individuals understand their overall orientation and form a correct spatial concept. Orientation signs can be perceived through various senses. Distinctive sounds and echoes, changes in ground texture or the structure of surfaces, pleasant or unpleasant smells, warmth from the sun or the direction of the wind, changes in elevation—all these must be integrated for effective O&M.

In essence, orientation points are consciously sought out by people with visual impairments, whereas orientation signs (such as noise or thermal changes) are often used subconsciously or for verification if the person with visual impairments loses direction (for instance, on the border between asphalt and stones).

Figure 9-1 presents examples of orientation points that are formed by corners of apartment buildings, while diverse orientation signs can be identified in (1) the border between asphalt and stones; (2) the noise (and thus direction) of traffic on the main street; and (3) auditory and thermal changes when the space beyond the buildings opens on the right.

The role of maps in O&M training

Establishing a correct idea of space (a spatial concept) is necessary for the truly independent movement of people with visual impairments. Acquiring such a concept is not easy, and this skill needs to be systematically developed, especially for people who have been blind since birth. Here, we face two challenges.

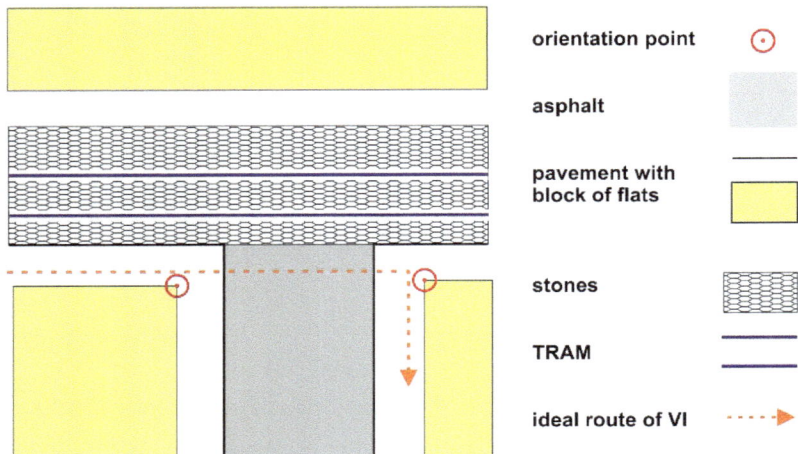

Figure 9-1. Example of the orientation points and signs in a basic urban environment.

The first challenge lies in the availability of tactile orientation maps and has traditionally been addressed by mostly manual production of orientation plans. Such plans were produced in a lengthy and costly manner and gradually became obsolete, as updating them was technologically complicated. With the advent of new technologies and especially with the advent of GIS, procedures have been implemented that allow for the quick preparation of up-to-date tactile plans. Based on a user-selected area of interest, these generate ready-to-print files, automatically creating map legends, and offer the ability to combine individual map sheets to cover larger areas.[5]

The second challenge is the significant time investment required to teach tactile map reading. This skill should be introduced as early as possible. In the following sections, we present two examples that have proved to work well.

Combining 3D models with 2D tactile maps

In a multiyear collaboration with an elementary school for the blind, we iteratively developed a tactile map reading set, combining a 3D model and a 2D tactile plan of the school's surroundings (figure 9-2). The model (figure 9-3) and plan (figure 9-4) were at the same scale and covered an identical area. The 3D model shows the basic layout of the buildings, including their height relationships. Being closest to reality, it is suitable for introducing spatial relationships to beginners. Later, it provides information that cannot be found in the tactile plan, especially about elevation. The same scale of the plan (with less distortion) helps refine the size relationships of

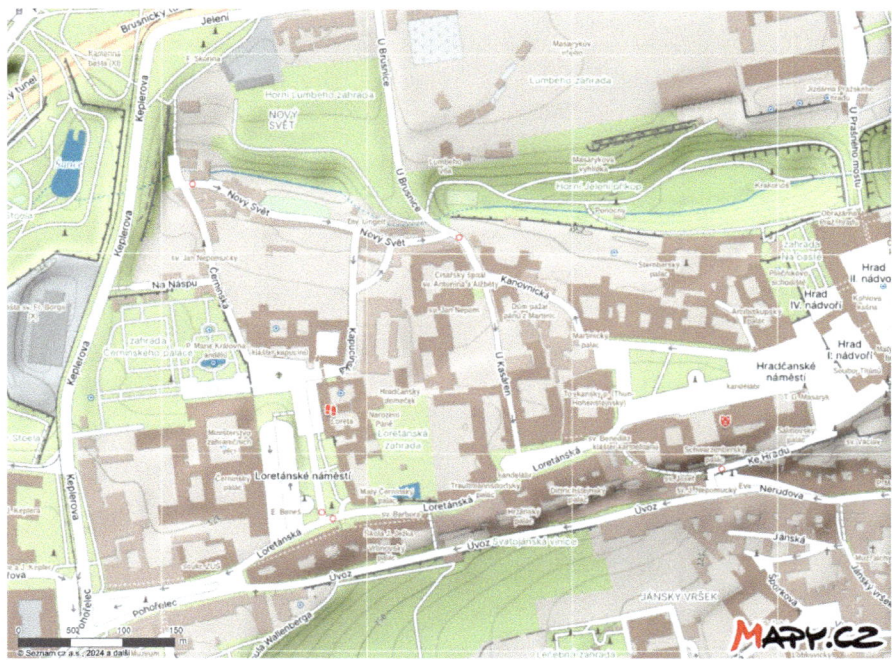

Figure 9-2. Map of the area depicted on a 3D tactile model of school surroundings (area approximately 800 × 700 meters).

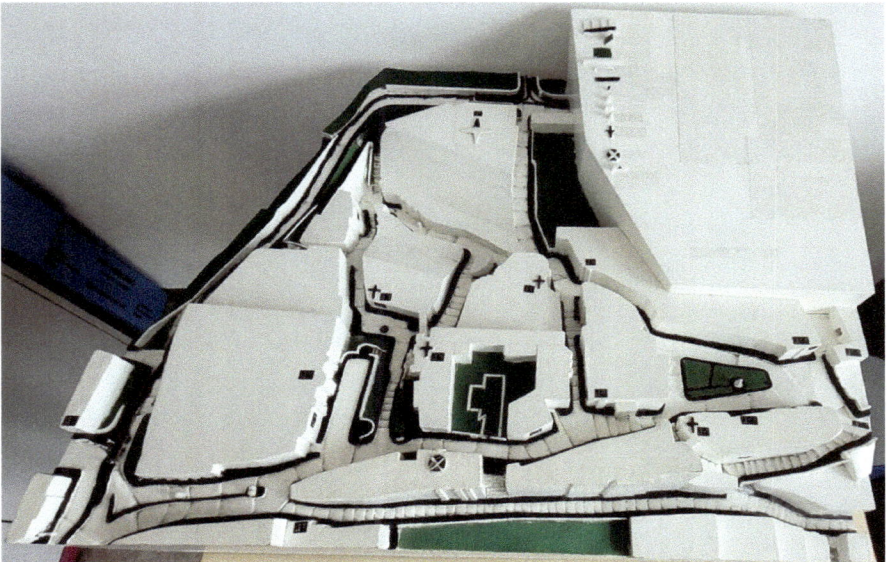

Figure 9-3. 3D tactile model of the school surroundings (scale 1:1000, exceeded 2.9 times).

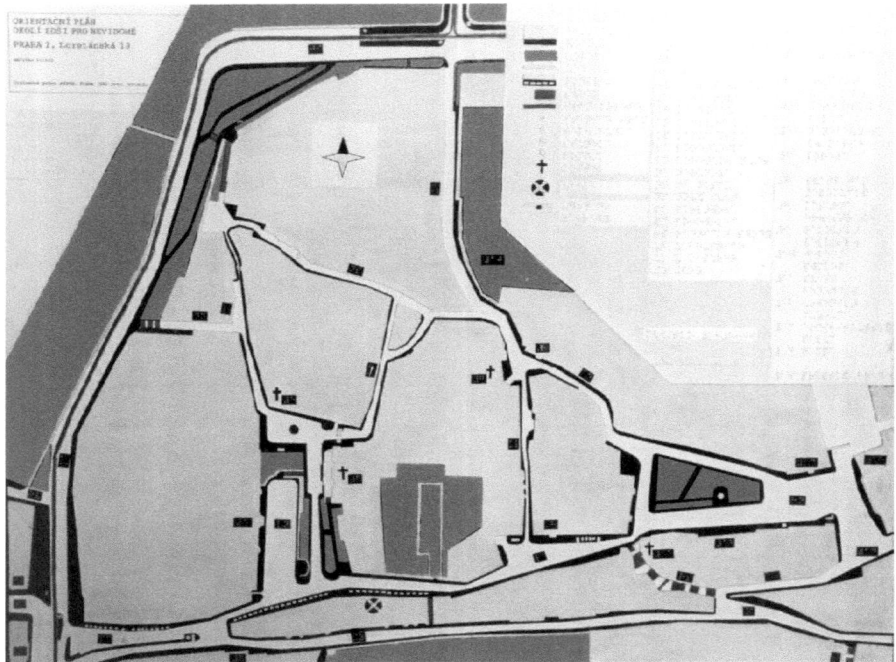

Figure 9-4. 2D tactile plan of the school surroundings (scale 1:1000).

individual objects (most evident in road widths). Originally, the model-plan combination was complemented with a schematic tactile plan providing a layout of the area on a smaller scale. However, it turned out that when using the model and plan on a regular basis, pupils acquired their own cognitive map of the school's surroundings, making the schematic plan obsolete.

Spatial navigation in unfamiliar settings

During a summer meeting of visually impaired youth from European countries, we tested how well participants were able to work with spatial information in an unfamiliar environment. We offered the participants a set of three maps differing in scale and taken from existing learning resources. The young people with visual impairments received a map of an auditorium, at scale 1:50 (figure 9-5), a map of the building with individual floors at scale 1:300 (figure 9-6), and a map of the building surroundings and the adjacent city center at a scale of 1:2700 (figure 9-7).

Participants in the auditorium were informed that the map of the building could be found in the auditorium with the help of the simple map of the room (see figure 9-5). They were asked to locate that room on the building map (see figure 9-6). Next,

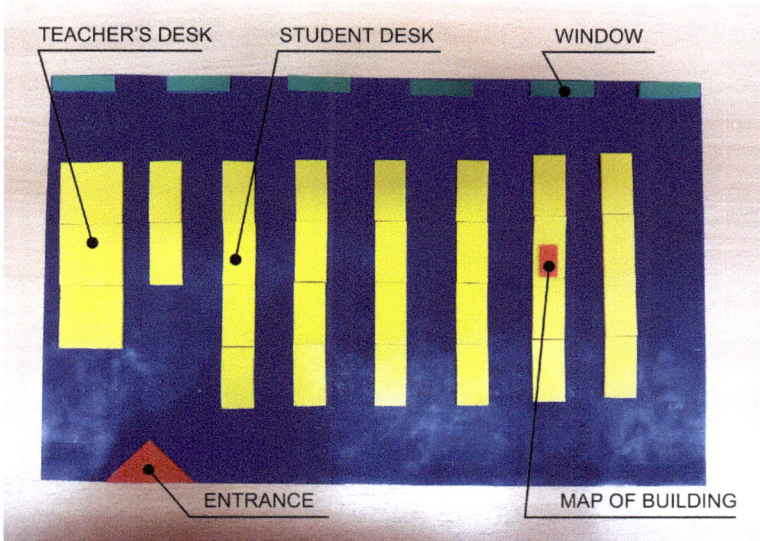

Figure 9-5. Tactile map of the auditorium (scale 1:50, dimension 10 x 9 meters, construction kit) with the position mark of the next map (map of the building).

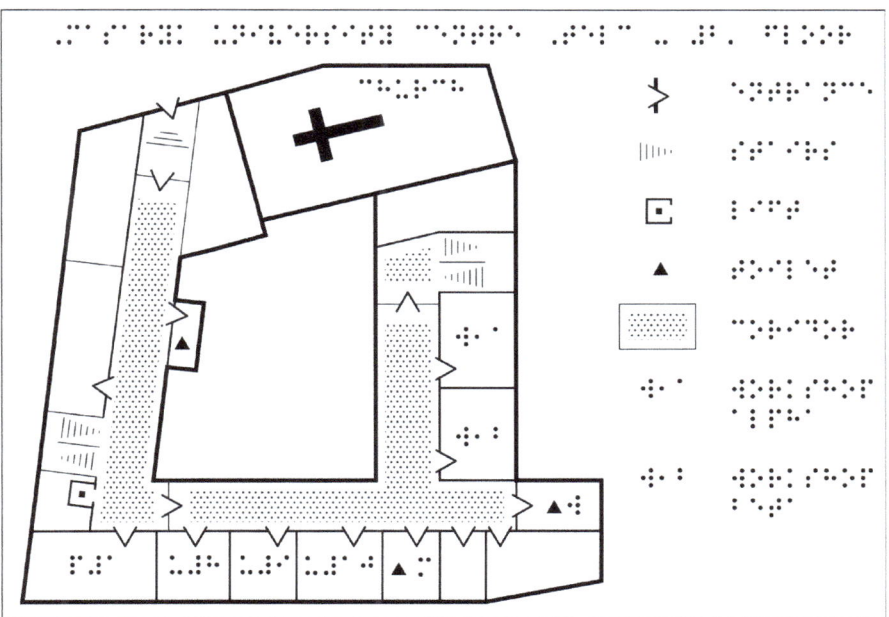

Figure 9-6. Tactile map of the first floor (A4, scale 1:300, swell paper) with a braille description of the classrooms in use.

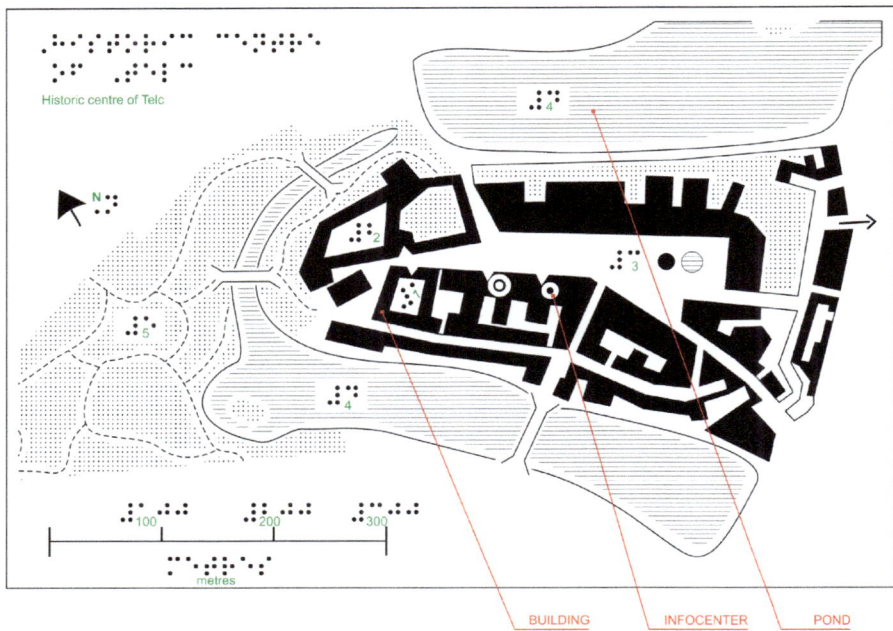

Figure 9-7. Tactile map of the city center (A4, scale 1:2700, swell paper) with a braille description of the main places.

using the building map, the participants had to find the route from their location (auditorium P1) to the workshop room (alpha). Once there, they gained access to the surroundings map (see figure 9-7), which had to be used to find the route to the information center. Finally, at the information center, participants received information about boat rentals on the local pond. Participants were, of course, expected to have mastered basic safe independent movement skills, and a successful completion of all these tasks rewarded them with not only a boat trip but, more importantly, the awareness that with the right sources of spatial information—maps—it is possible to move even in an unfamiliar environment.

Will navigation apps replace tactile maps?

Often, we encounter the opinion that tactile maps are obsolete. The orientation challenges of people with visual impairments are assumed to be completely solved by mobile navigation apps. But rather than competing, maps and apps are complementary. Apps greatly facilitate navigation, especially in determining one's location and the direction and distance to a given destination, which is difficult, if not impossible, to obtain from a tactile map. A tactile map, however, provides comprehensive information about the entire space in which a person is moving. Although the navigation app takes us "exactly" to our destination, only a map can tell us what the surroundings look like.

Looking ahead, combining navigation with additional image analysis, either by a remote vision assistant (for instance, Be My Eyes app) or AI-assisted image analysis (for instance, Envision AI or Biped NOA), seems promising; such tools could provide real-time visual information to people with visual impairments, helping them navigate unfamiliar environments with greater ease and independence. Although tools like this will be helpful in increasing the situational awareness of people with visual impairments, they aren't expected to quickly abstract an overview of an entire area. That's the role and unique function of a tactile map.

Practical insights for work with tactile maps

For many, maps are an essential part of their everyday professional work; for others, they are a useful source of information, knowledge, and entertainment. Some do not seek out maps or use them at all. The same applies to people with visual impairments.

Tactile maps are not always the best solution

In certain situations, other sources of information are far more appropriate for people with visual impairments than maps. Take, for example, the schematic maps commonly displayed in zoos to indicate the natural habitats of animals in the world. Such maps can contain multiple colored borders, indicating the particular habitats of animals. With only one glance, sighted people can get a clear idea of where a specific animal lives. Although it is certainly possible to create a tactile map representing the same information, a tactile map may not be the most efficient medium to communicate it. People with visual impairments have to invest disproportionately to obtain the required information by touch. First, they need to orient themselves in a schematic map of the world, then search systematically for the distinct surfaces indicating the specific habitat, and finally place those habitats in the context of the continents. The accuracy and degree of certainty from the information obtained is very low. One single audible or braille printed sentence could resolve the situation much more quickly, accurately, and efficiently.

Overviews must be built, and context is required

Seeing maps is fundamentally different from feeling them. A sighted person can glance over the map to find out what territory is shown and what elements are used for the map's content. In a few seconds, the person is basically oriented to the map in front of them and can go about finding the necessary details. For a person with visual impairments, such independent exploration of a map is time-consuming and requires a relatively significant amount of energy, not necessarily leading to a correct understanding of the map.

To optimize users' understanding of a tactile map, certain rules are helpful. People with visual impairments need sufficient time to become familiar with the tactile map in detail, preferably in a comfortable place, at a table, without distractions. And because even experienced map readers do not easily find their way on a map, maps should always be accompanied by supplemental information given in advance to people with visual impairments to familiarize the user with basic information about the map before they read it. This approach tries to mimic the first-glance impression of sighted people for people with visual impairments and includes information on the title, a brief description of the territory that the map shows, and basic expressive

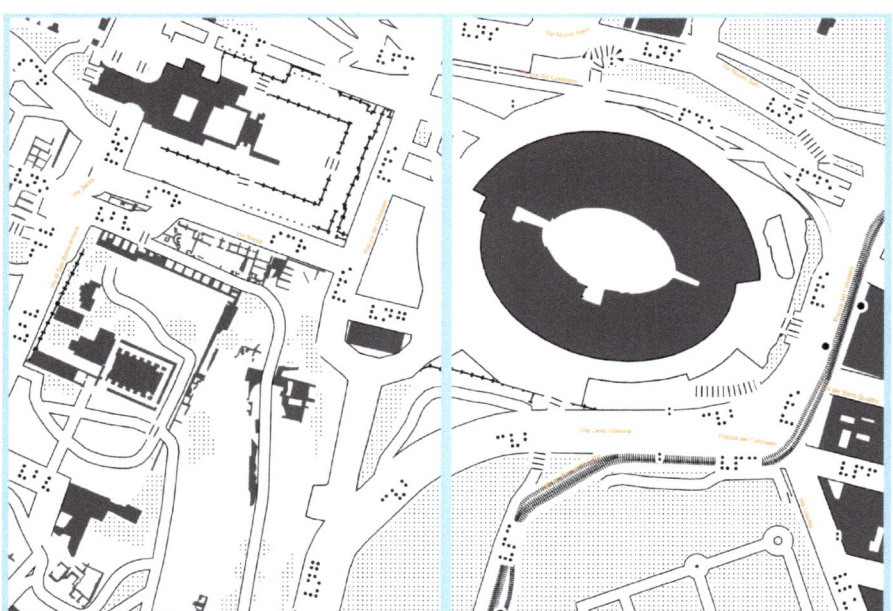

Figure 9-8. Tactile map sheets j3880_u5654 and j3881_u5654 from tactilemaps.eu (each of the maps should be printed on A4 swell paper, scale 1:1300).

A virtual walk around the Colosseum

1. The tactile map of the Colosseum area consists of two connected map sheets, left and right. The scale of the map is 1:1300, which means that one centimeter on the map corresponds to 13 meters of real territory. The distinctive ellipse of the Colosseum is located on the right sheet.
2. On the inner perimeter of the Colosseum, the entrances through the amphitheater are visible, narrower on the west and east sides, wider on the south side.
3. Pass through the imaginary southern entrance out of the Colosseum and follow its outer perimeter closely around. Note that the plan of the amphitheater is wider on the north side than on the south side.
4. Continue south from the south side of the Colosseum and find the abbreviation CVB, which indicates the wide street Via Celio Vibenna. This is a road partially encircling the Colosseum from its south to east side.
5. Take Via Celio Vibenna east, keeping to the left edge of the street. As the street begins to curve north, just after the next CVB abbreviation, you will bump into steps at the left edge.
6. At the stairs, cross to the opposite side of the street, just above the PDE, Piazza del Colosseo.
7. Continue along the right side of the street a little to the north. You will come to a small round sign indicating a tram stop.
8. On the right of the tram stop, there is a double line indicating the track. Imagine that you are getting on a tram that has come from the north and is heading down the track to the south.
9. You first pass several buildings on the left in the direction of travel. As the tram turns west, more greenery will appear.
10. The track turns south again, with greenery on both sides. The line is interrupted in several places by the abbreviation VPC, Viale del Parco del Celio.
11. At the point where the tram line leaves the map, there is another tram stop to the right of the track. At this stop, turn around and follow the track back to the original stop.
12. From the tram stop, take the familiar Via Celio Vibenna (CVB) back to the left, heading southwest. The street will lead you to the left side of the map.

The virtual trip can continue to the second map sheet (facing page, left map).

elements (sometimes the whole legend is included), but mainly the basic location of the most important objects on the map. Such information could be provided by a sighted assistant or, when such assistants are not available or when the map is legible and enhanced with audio descriptions, provided in the form of an introductory sheet.[6] This serves as a "virtual walkthrough" of the map, guiding a reader through the most important map elements, helping people with visual impairments generate a mental image of the mapped area. Although this is time-consuming, we have found that it improves the understanding of individual map elements and the spatial context of the displayed area. As an illustration, we have included a virtual walk around the Colosseum in Rome, accompanying the map in figure 9-8.

Conclusion

The availability of tactile maps is increasing significantly, which is, of course, a positive development. But the question remains whether the skills of people with visual impairments to read these maps are also increasing—or better still, whether the number of skilled users of tactile maps is increasing. In our experience, much more could and should be done with tactile maps in practice. There is still a relatively large group of people with visual impairments who come to maps much later than they need to. And some of them do not encounter them at all or only very superficially.

Just as mastering orientation is essential for a truly independent life, so tactile maps are essential for the overall development of general knowledge and a proper understanding of space, which is not limited to one's immediate surroundings. Maps of schools and historical parts of cities, plans of parks and museum exhibitions—these and more can be used in teaching map-reading skills. These skills are not only essential in O&M but also in the everyday lives of people with visual impairments, preventing their social exclusion. Through a proper combination of tactile models and maps, we can help people with visual impairments perceive, understand, and explore the surrounding area, town, country, and, indeed, the world. In addition to the development of methods of production, we must not forget the development of effective methods of working with maps and making them accessible to people with visual impairments.

Further reading

On orientation and mobility of the visually impaired

Blash, B. B., W. R. Wiener, and R. L. Welsh. *Foundations of Orientation and Mobility*, 2nd ed. AFB Press, 1997.

Cratty, B. J. *Movement and Spatial Awareness in Blind Children and Youth.* Charles C. Thomas, 1971.

Hoover, R. E. "The Cane as a Travel Aid." In *Blindness*, by P. A. Zahl. Princeton University Press, 1950, 353–65.

Jesenský, J., et al. *Study Materials for Spatial Orientation and Independent Movement for the Visually Impaired* (in Czech). Prague: SI v ČSR, 1978.

Wiener, P. *Orientation and Independent Movement of the Visually Impaired* (in Czech). Prague: AVICENUM, 1986.

Wiener, P. *Orientation and Mobility of the Visually Impaired* (in Czech). Prague: Rehabilitation Institute for the Visually Impaired, Charles University in Prague, 2006.

Wiener, P., and P. Červenka, eds. *Report from the Fifth European Seminar on Education of O&M Instructors.* Prague: Charles University, 1995.

Wiener, W. R., R. L. Welsh, and B. B. Blasch. "Foundations of Orientation and Mobility." In *Instructional Strategies and Practical Applications*, 3rd ed., vol. 2. AFB Press, 2010.

On tactile map creation

Edman, Polly K. *Tactile Graphics.* American Foundation for the Blind, 1992.

Eriksson, Y., J. Gunnar, and M. Strucel. *Tactile Maps: Guidelines for the Production of Maps for the Visually Impaired.* Enskede: The Swedish Library of Talking Books and Braille, 2003.

NSW Tactual and Bold Print Mapping Committee. *A Guide for the Production of Tactual and Bold Print Maps*, 3rd ed. Surry Hills: The NSW Tactual and Bold Print Mapping Committee, 2006.

On tactile map usage

Bentzen, B. L. "Orientation Aids." In *Foundations of Orientation & Mobility*, by R. L. Welsch and B. B. Blasch. AFB Press, 1980, 291–355.

Červenka, P. *Maps and Orientation Plans for the Visually Impaired: Creation and Usage Methods* (in Czech). Prague: AULA, 1999.

Červenka, P. "Ten Years' Experience Using the Tactile Models and Plans." In *Proceedings of the 11th International Mobility Conference*. Stellenbosch, South Africa, 2003.

Červenka, P., M. Hanousková, K. Břinda, R. Seifert, and P. Hofman. "Blind Friendly Maps: Tactile Maps for the Blind as a Part of the Public Map Portal" (Mapy.cz). In *Lecture Notes in Computer Science: Computers Helping People with Special Needs*, by C. Buhler, P. Penaz, and K. Miesenberger. Switzerland: Springer Verlag, 2016, 131–38.

Acknowledgments

The first section, "Micro-orientation and macro-orientation," is indebted to the longitudinal research and practice of my former colleague, Pavel Wiener. We gratefully take his research further by adding years of collaborative work with O&M trainers and blind clients of all ages.

Pavel Wiener (1956–2021) graduated from the Faculty of Education of Charles University in the field of special pedagogy—education of the visually impaired. He worked for more than 30 years in the field of orientation and mobility of the visually impaired. He was the author of a new teaching method of spatial orientation education, *lege artis* (recognized by the Ministry of Education of the Czech Republic). He was also founder and longtime director of the Rehabilitation Institute for the Visually Impaired at Charles University in Prague (1994–2008). My teacher, colleague, and friend.

Pavel Wiener.

About the author

Petr Červenka is a researcher and orientation and mobility trainer at Masaryk University, Support Center for Students with Special Needs, Teiresiás, Czech Republic. After graduating from Charles University in Prague with an MSc in geography and cartography, he qualified as an orientation and mobility trainer and instructor of the visually impaired (1993, 1998). From 1994 to 2008, he participated in the preparation of orientation and mobility trainers at the Rehabilitation Institute for the Visually Impaired, Charles University, Prague. Červenka is a member of the Blind Friendly Maps project team, which has produced a public web service for the automated generation of tactile maps (tactilemaps.eu). He's also a music lover (former sax player) who enjoys mountains and traveling.

Petr Červenka.

Notes

1. Jesenský, J., et al, *Study Materials for Spatial Orientation and Independent Movement for the Visually Impaired* (in Czech), (Prague: SI v ČSR, 1978).
2. Wiener, P., and P. Červenka, eds., *Report from the Fifth European Seminar on Education of O&M Instructors* (Prague: Charles University, 1995).
3. Wiener, P., *Orientation and Independent Movement of the Visually Impaired* (in Czech) (Prague: Avicenum, 1986).
4. Hoover, R. E., "The Cane as a Travel Aid," in *Blindness*, by P. A. Zahl (Princeton University Press, 1950), 353–65.
5. One such system that allows the preparation of detailed maps for teaching spatial orientation is the web application tactilemaps.eu; cf. Červenka et al. (2016) and bfmaps.org.
6. NSW Tactual and Bold Print Mapping Committee, *A Guide for the Production of Tactual and Bold Print Maps*, 3rd ed. (The NSW Tactual and Bold Print Mapping Committee, 2006).

Case study

A collaborative approach to tactile mapping in the Netherlands

Jolijn Jansen

Many product designers design for the average user or use a statistical rule to accommodate the middle 90 percent of users, which means that people with a disability are often left out of the design process. To address this situation and provide people with visual impairments with better access to geographic data, we undertook an interdisciplinary study, combining the expertise of people with visual impairments (experts by experience) with the knowledge and skill of four organizations. Kadaster provided expertise in geographic data, the Dutch Accessibility Foundation contributed user research and applied advice, Dedicon offered knowledge of tactile reading and map production, and Esri® Nederland provided the necessary software solutions.

From the start, the end users were involved in a user-centered design process, with five iterative rounds of user research. In these rounds, users cocreated the map design and symbology by giving essential input to determine the goal, brainstorm possibilities, review the pilot versions, and eventually evaluate the product.

As a person with visual impairments, Arend Jan van Dongen participated in the project. He describes a real-world example of the map's utility: "On one of the test days, I received a beautiful map of my neighborhood in Vught. It was only then that I realized how my surroundings were laid out. When I walk through the alley and past my house and cross the street where I live, I can continue walking in another alley on the other side. But every time I did that, I ended up in the wrong place, and I couldn't understand why. With the map, I discovered that the alley next to my house is not directly opposite the alley on the other side. I always thought I was crossing the street diagonally!"

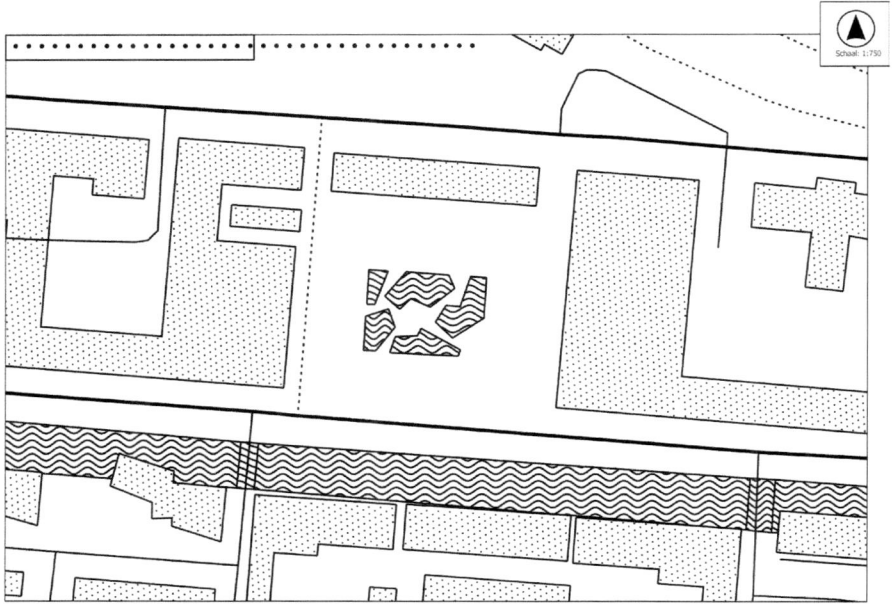

First draft of a tactile map of Lubeck Square, Zwolle, the Netherlands. Image courtesy of Dutch Tactile Map Project.

The first round of the design process used focus groups to establish the needs and wishes of end users. We learned that tactile maps are mostly needed to supplement mobile navigation apps to provide a bigger "geographic picture" because those apps don't inform the user about details such as the sharpness of a bend or the spatial relationships between locations within a larger area. Tactile maps also require less detail than regular maps and need clear orientation points, such as the houses of friends or locations known to the people with visual impairments for a specific sound or smell (for example, a bakery). Most people with visual impairments use public transport to get around, making this an important theme for the maps. Based on this feedback, we designed a draft tactile map, which was reviewed by an additional seven users. They suggested potential improvements, such as more distinguishable symbols, additional spacing between area textures as well as between area textures and roads, and a tactile scale indicator.

The second round focused on tactile symbology for monochrome swell paper maps. Based on the existing literature and best practices, we designed point, line, and area symbols and conducted a tactile discrimination test to determine the most distinguishable and most contrasting combinations for each symbol type. The

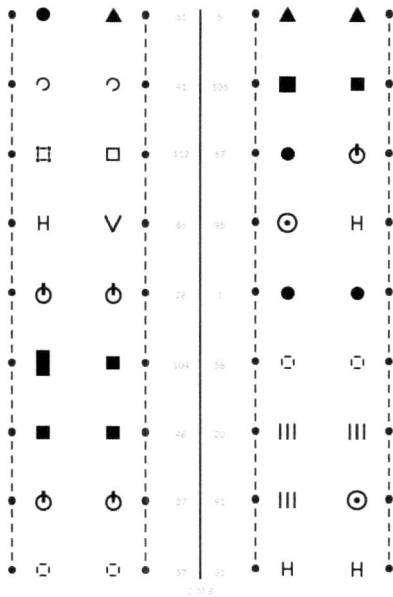

One of eight sheets from the tactile discrimination test. Image courtesy of Dutch Tactile Map Project.

participants were asked independently to assess within three seconds the (dis-)similarity of two symbols for a set of 60 to 144 combinations, depending on the number of tested symbols (6 to 9 symbols). The test was based on signal detection theory, which explains how people determine the presence or absence of a stimulus. From these tests, we deduced a well-distinguishable symbol set of seven point symbols, four line symbols, and four area symbols out of 21 tested symbols.

The third round of research focused on decreasing the cognitive load of map reading. Tactile map reading requires an additional step for people with visual impairments, who must work from the map details to construct a mental overview of the map. This additional step makes the cognitive load much higher and, in our research setting, even caused some participants to abandon the task early. The feedback of two focus groups showed that a two-factor approach can be applied to relieve the cognitive load. First, the people with visual impairments will get a short map description (150 words maximum). This describes the contents of the package: three complementary maps and a legend. The text then describes the most important objects on the maps, going clockwise from the most pronounced orientation point. The map information is offered in three maps that build up from minimal to

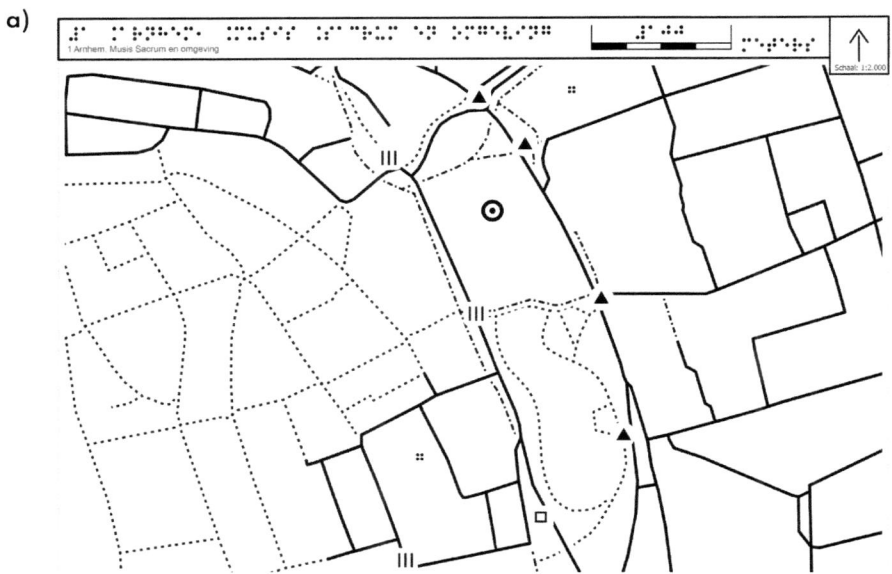

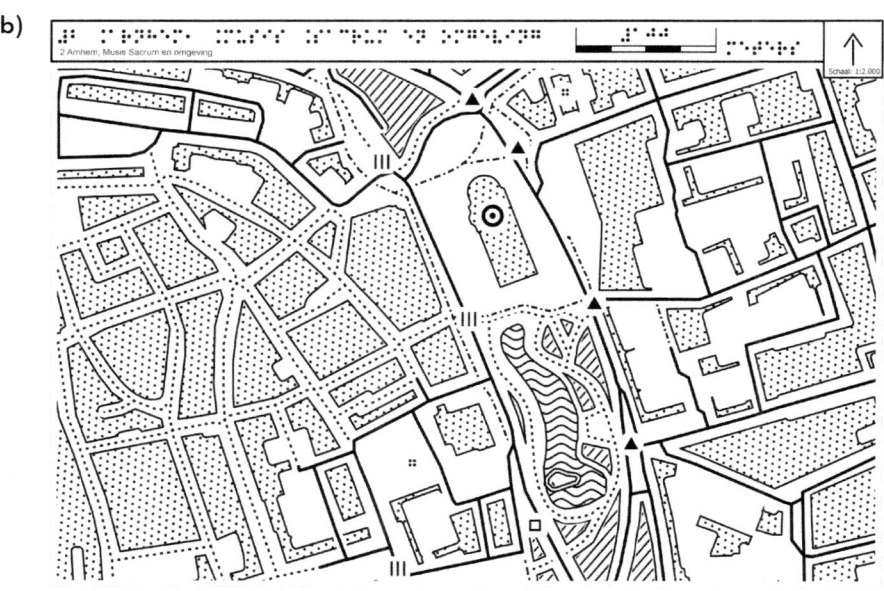

Three maps included in the map package, which build in complexity. The first map (a) is Musis Park and surroundings. In (b), the area symbols are filled in, and in (c), the map is zoomed in, and the park can be recognized by its oval shape. Image courtesy of Dutch Tactile Map Project.

comprehensive geographic information. By layering information density and adding a description, this approach enables people with visual impairments to build an overview more easily, leading to higher information retention.

By the fourth research round, we had established the map package, symbol design, and a functioning map that were ready for refinement. We organized individual research with eight participants in which they would explore the map and answer in-depth questions, such as how they would interpret this intersection of lines or follow that footpath through the city center. Based on this feedback, we changed spacing and line thickness to increase the differentiability of map elements. We also added color to aid people with less severe visual impairments, as

Example of one tactile map with color added. Image courtesy of Dutch Tactile Map Project.

well as sighted people. Therefore, the map is not only suitable for people with visual impairments but can be used by all.

The user-centered and iterative design process was key to the development of a tactile map package that caters to the needs and wishes of end users. It has a legible map design and symbology and helps users orient themselves by gradually providing more detailed spatial information, enabling them to build a mental map.

As Arend Jan van Dongen explains, "I'm involved as one of the experts by experience, and I find it very valuable that I am fully involved in the process. Over the years, I've often experienced people making decisions for me, and how they would then miss the mark. But in this project, we're truly working together to design tangible maps. From everyone's expertise. And I'm happy to contribute from my practical experience."

By involving the end user throughout the design process, we not only created tangible tactile maps but ones that suit the wishes and needs of our target audience, making tactile maps more usable, understandable, and user-friendly for people with visual impairments.

Case study

Tactile world thematic map

Young-Hoon Kim

In recent years, the government of South Korea has implemented various policies to support marginalized social groups. Nevertheless, people with visual impairments continue to experience challenges in accessing information, primarily because of the prevalence of image- and video-based content across digital platforms. This situation presents substantial barriers to learning, exacerbating disparities in social participation, adaptation, and access to geographic data.

In 2015, the National Geographic Information Institute (NGII), South Korea's national mapping agency, developed a tactile world map specifically designed for visually impaired students. This map, which includes major cities and oceans, was distributed to schools serving people with visual impairments throughout the nation, establishing a foundational representation of global geography. A teacher at a school for individuals with visual impairments emphasized the significance of tactile maps, stating that they enhance educational accessibility and foster equal learning opportunities. Educational institutions specializing in special education use this tactile map in their world geography curricula, enabling students with vision impairments to engage more effectively with the subject. The original world map has been modified to incorporate a series of additional thematic maps, enhancing students' access to and comprehension of world geography.

Before selecting topics for thematic tactile maps, the NGII conducted a comprehensive analysis of world map topics found in the national atlas and in textbooks for primary through high school. From this, we learned that the representation of physical and cultural elements and basic global geography feature prominently and that thematic maps incorporating additional information are necessary to provide greater geographic comprehension and spatial understanding for people with visual impairments. This analysis also enabled the researchers to identify key subjects, leading to the development of thematic tactile maps tailored to each selected area, using data from the NGII's geographic information repository.

The tactile world maps serve as specialized educational resources for students with vision impairments, with 18 different types distributed to schools for the visually impaired across South Korea. The tactile maps were created using UV printing technology, and feedback was collected from students at nationwide schools for the visually impaired. The evaluation process included usability testing and satisfaction surveys. Overall, the tactile maps received high satisfaction levels from participants; key feedback indicated a need for more prominent symbols on the map, especially for the lines representing meridians and parallels—latitude and longitude, which should be more distinguished. Additionally, the representation of landforms should be more explicit, with a clear differentiation between continents and oceans.

The NGII has remained committed to providing these maps to enhance geography education for people with visual impairments. In educational settings, these tactile maps serve as effective supplementary materials, fostering the development of geographic knowledge and spatial awareness. The maps not only aid in a classroom setting but also help engage parents in their children's education. A parent of a child with visual impairments observed that tactile maps helped them effectively teach their child geographic concepts, even without any prior knowledge of braille.

The growing demand for tactile maps underscores the need for ongoing production and distribution of diverse tactile resources addressing various global themes. To address these needs, the NGII intends to expand its scope by developing additional thematic tactile maps related to a wide range of global geography topics, such as a tactile globe, audio services, and digital tactile maps. This will enhance the resources available to students with vision impairments in educational and everyday contexts. For example, one student stated that using tactile maps enabled them to navigate unfamiliar places more confidently, significantly increasing their independence.

Over the past seven years, South Korea has accumulated considerable expertise in producing tactile maps, involving collaborations among its national mapping agency, private-sector entities, and academic institutions. The private sector's advancements in tactile map technology have further contributed to the increasing use of these maps in geography education at schools for the visually impaired. We anticipate that tactile world maps will be increasingly prominent in advancing geography education and equal learning opportunities for students with vision impairments.

NGII list of 18 types of tactile world maps for people with visual impairments

1. The Five Oceans and Six Continents
2. Latitude, Longitude, and the Equator
3. The Antarctic Region (Antarctica)
4. Distances Between the Republic of Korea and G20 Member States
5. The Arctic Region
6. Major Rivers of the World
7. Prominent High-Altitude Mountain Ranges
8. Major Plateaus of the World
9. Globally Prominent International Airports
10. Major Desert Systems Worldwide
11. Principal Mountain Ranges of the World
12. Global Ocean Currents and Their Flow Patterns
13. Global Volcanic Zones and Seismic Belts
14. Global Climate Classifications
15. Global Major Landforms
16. Global Tectonic Plate Structures
17. UNESCO World Heritage Sites
18. Comparison of the Sizes of Major Countries

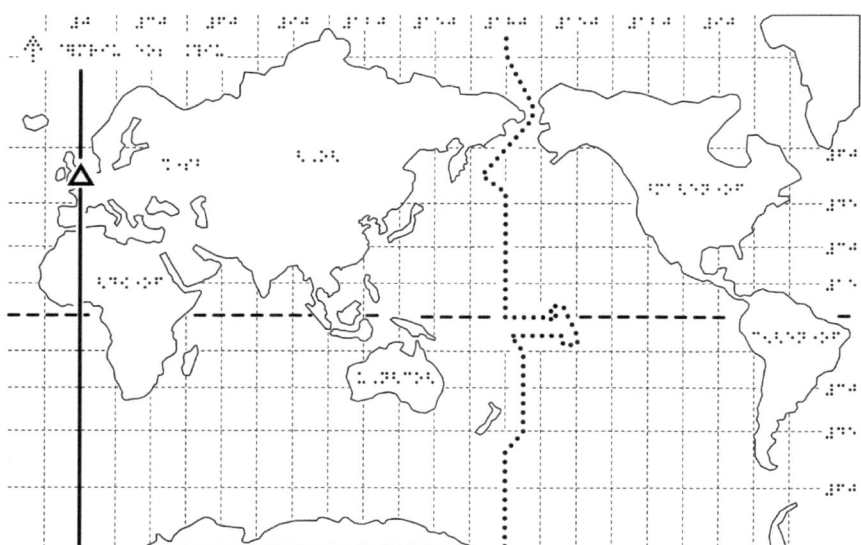

A world map in braille with a legend at the top. Image courtesy of National Geographic Information Institute.

Tactile world thematic map 199

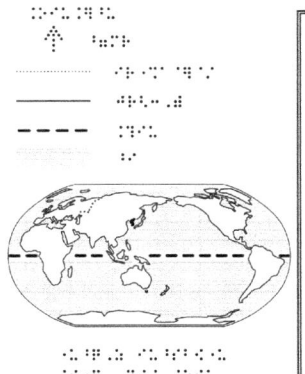

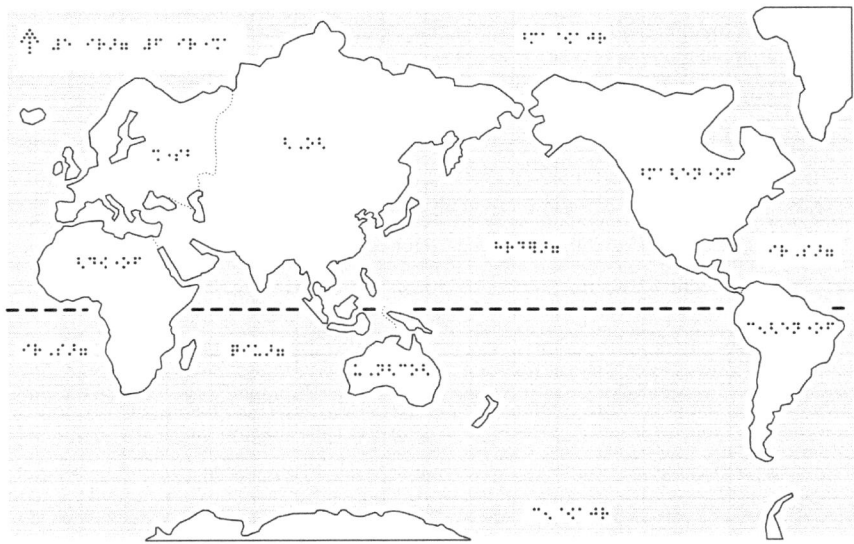

Thematic tactile map of the five oceans and six continents, with a legend at the top.
Image courtesy of National Geographic Information Institute.

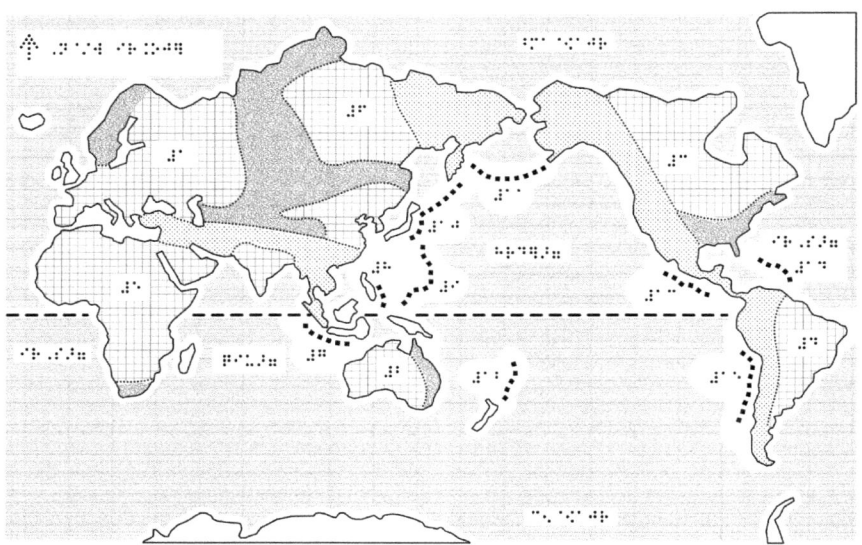

Thematic tactile world map focusing on the Eastern Hemisphere, with landmasses and oceans. Image courtesy of National Geographic Information Institute.

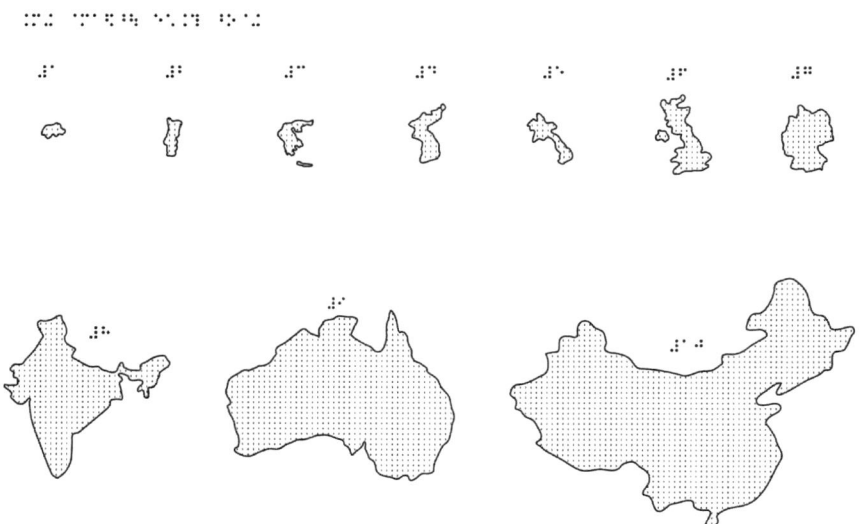

A comparison of the sizes of several countries, including, *left to right*, India, Australia, and China. Image courtesy of National Geographic Information Institute.

Part V
Reliable output

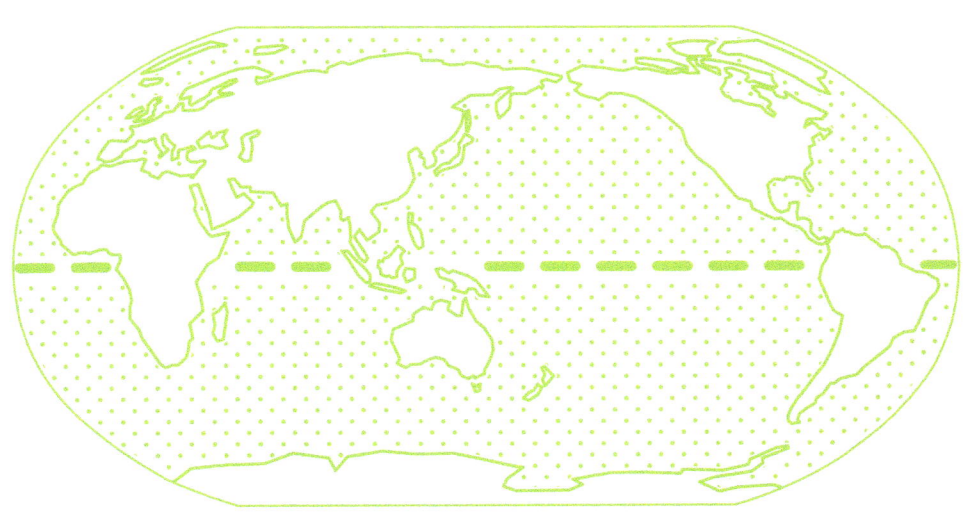

Personal story

We drop things at the same rate

Dorothy Atieno Lensa

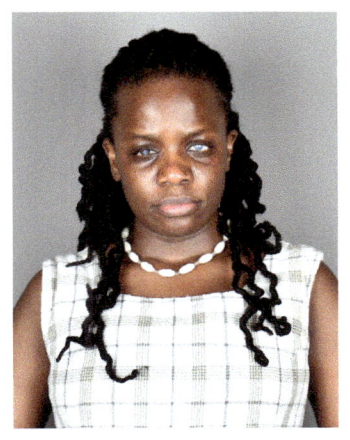

Dorothy.

My condition started with low vision when I was about nine years old, and it slowly deteriorated to blindness by the time I was 13. I was lucky that I did not have to go to a special school because I lived in an area of Nairobi, Kenya, with an integrated school system that served children without disabilities alongside a small group of blind and low-vision learners like me.

It was difficult transitioning to using braille in class, though. Early on, I had a hard time reading and interpreting diagrams and graphs with my fingers because my school used a low-quality Thermoform machine to create braille printouts. Even though the graphs and diagrams printed in braille were already oversimplified, the machine my school used typically created unintended gaps in the raised dots, which impeded the printouts' tactile sensitivity. I remember being very worried about the graphs and diagrams that would appear in my final exams, only to find that they were clear and of high quality. I wondered why such high-quality materials were made available to us only once, at the end of our high school careers.

Both my high school and university had computers with screen-reading software that was supposed to make learning easier. I could not use the computers, however, because access to them was severely limited at both institutions. I did not understand why these schools could not provide us with computers that we could use on a regular basis. When I asked about this, staff at the university told me they were worried that we would drop and damage the expensive machines. But we never dropped or damaged our heavy and just-as-expensive braillers!

People with visual impairments drop things at the same rate as sighted people. If sighted people aren't known for dropping and damaging their machines, why would we be?

Chapter 10

Accessible media

Radek Barvíř, Alena Vondráková, and Jan Brus

In September 2009, as I started my PhD, I joined a two-year running project aimed at creating "modern types" of tactile maps. At the time, "modern" meant tactile maps in color, 3D modeled, and, ideally, with some audio features. Despite our confidence in the available technology, expertise, and ideas, the process faced unexpected challenges.

The only printer for individual prototyping used gypsum powder, producing fragile and expensive models. User testing revealed insufficient durability and flawed map symbology due to the inadequacy of tactile-graphic parameters used in plastic tactile maps at that time. The project was supposed to end in 2010 with prototypes of multimedia tactile maps, but by late 2009, we could only produce static maps, not the planned multimedia versions. Additionally, all our theories on how this should work (for instance, using modern MP3 players) failed.

Seeking help from computer scientists, we received a rudimentary prototype built with a fire alarm, motherboard, and speaker, leaving us to complete it independently. I purchased hardware components and crafted wooden boxes. With drills, glue, soldering irons, and, most importantly, team effort, we assembled prototypes of multimedia tactile maps within two weeks—entirely impractical for real-world use. Although the project objectives were formally met, our vision was not.

We did not give up. In 2014, with Jan Brus and Radek Barvíř, we developed a viable solution: TouchIt3D, combining low-cost 3D printing with touch screen tablets. This technology has addressed earlier issues and continues to evolve, improving accessibility and usability. For more than 10 years, we have been developing and implementing this technology in practice, with the enthusiasm of its users as our greatest reward.

—Alena

Which one to choose?

As we'll explore, the production of tactile maps offers a rich variety of techniques, each with its strengths and limitations. From handcrafted methods and swell form machines to 3D printing, embossing, and thermoforming, each method has its unique advantages, depending on the application. The choice of technique often depends on factors such as the level of detail required, the durability of the final product, budget considerations, and whether visual elements, such as color or texture, will be integrated. There is no one-size-fits-all solution. What works best for a simple educational tool might differ from what is ideal for an outdoor navigation aid. As you reflect on these methods, consider the goals of your tactile map: Who will use it? Where will it be used? What features are most important? How many copies do you need? These questions will guide you toward the technique that aligns perfectly with your vision, challenging you to think critically about how to turn accessibility into reality.

Evolution of tactile map creation

Designing tactile maps for individuals with severe visual impairments involves a challenging blend of creativity, technology, and sensory understanding. From the earliest engravings on clay and bone, the evolution of media in tactile map creation reflects technological advancements and an increasing emphasis on accessibility and usability. Progressing from simple manual methods to advanced multimedia systems, this evolution has been marked by continuous improvements in precision, accessibility, and user experience. Today, these advancements have expanded the role of tactile maps from essential orientation tools to comprehensive educational and navigational resources.

Manual methods

Most manual techniques have served as the foundation for the future mechanized production of tactile maps. Although manual creation is suitable for producing individual custom-made maps, mechanized processes enable mass production. However, the role of manually created maps remains significant, as they allow for quick creation and provide an effective representation of the required spatial information. A wide range of materials are employed in the manual creation of tactile maps, including palm drawings, wood or paper-based techniques, fabric-based maps, clay and wax models, and synthetic materials.

Palm drawing is one of the oldest and simplest methods of conveying spatial layouts. This technique, primarily used to communicate directional and positional relationships to people with visual impairments, involves the informant tracing spatial arrangements on the recipient's palm or hand using their finger while verbally explaining the meaning of each symbol. The method relies on direct, tactile interaction, allowing for immediate clarification and feedback from the user. This makes it particularly effective for conveying simple spatial relationships in real time. Despite its advantages, palm drawing has notable limitations. The information conveyed is ephemeral, existing only in the recipient's memory, preventing repeated verification or detailed spatial analysis. Also, the method is constrained by the size and shape of the human hand, making it unsuitable for communicating complex or large-scale spatial layouts. Although palm drawing remains a valuable and immediate tool for basic spatial communication, it has been largely superseded by more permanent and scalable manual tactile map creation methods. These methods incorporate materials and techniques designed to address the limitations of palm drawing, providing more durable, repeatable, and detailed representations of spatial data.

Paper-based techniques are among the simplest and most accessible methods of tactile map creation. Paper is inexpensive, widely available, and easy for most people to work with, making it the most common type of manually created tactile map (figure 10-1). Layers of paper can be glued together to create a 3D relief. String, thin wire, or adhesive materials can be attached to the surface to represent boundaries, paths, or other linear features. Patterns can be manually pressed or embossed into

Figure 10-1. Pinpricking is one of the simplest methods of creating tactile graphics on paper. An ordinary pin and plain paper are enough to create simple tactile maps. Institute of Special Education Studies, Palacký University Olomouc.

the paper using a stylus or other pointed tools. This method is cost-effective but requires skill and precision to produce clear, recognizable symbols. Although particularly effective for basic spatial representations, paper-based tactile maps are less durable and more prone to damage compared with other materials.

Fabric-based tactile maps can also be created relatively easily with common household tools. Fabric offers flexibility and durability, making it a popular medium. Raised features are created by stitching or embroidering lines and symbols on the fabric. Different textures and thread types can represent various geographic elements. Layers of fabric or other materials, such as felt or leather, can be used to create distinct regions or topographic features. In the archives of the Perkins School for the Blind, there is a preserved tactile map of South America from approximately 1900. This map uses pins of varying sizes to represent the basic geography of the continent.

Clay and wax are highly malleable materials that enable detailed, three-dimensional representations. Clay can be shaped into raised features to depict terrain, providing a tactile representation of geographic elements. Wax sheets, on the other hand, allow for symbols to be engraved or pressed directly into the surface, offering a durable and versatile option.

Wood has historically played an important role in tactile map creation because of its durability and accessibility. Different layers of wood can be assembled to represent tactile graphics, such as elevations or distinct map features. Another method involves engraved wooden maps, where wooden boards are carved or engraved to create raised lines and symbols. Wooden-based tactile maps are highly durable, but their creation requires skilled craftsmanship and specialized tools. An example of using wood in tactile graphics creation is the so-called pegboard. This board has predrilled holes into which small wooden pegs can be inserted. Using this method, both simple and more complex patterns can be created (figure 10-2).

Introducing synthetic materials has greatly expanded the possibilities for manual tactile map creation. Synthetic materials offer considerable versatility, along with various types of plastics that can replace paper or fabric. For instance, hot glue guns, commonly used for decorative purposes, allow heated plastic to be applied as a liquid to create relief drawings on a surface. These designs can incorporate multiple colors or layered patterns through gradual application. Similarly, synthetic paints and varnishes enable the creation of extensive tactile graphics. By applying and drying successive layers, users can produce complex and detailed representations with virtually no limitations. This principle is also used by 3D printing or

Figure 10-2. With the use of wooden pegs and a pegboard, various shapes can be created. These temporary tactile maps cater to the immediate need for spatial representation. The color of the pegs helps individuals who rely on residual vision for recognition. Institute of Special Education Studies, Palacký University Olomouc.

advanced printing technologies that apply a varnish to paper and create a combination of printing and a colorless tactile layer (figure 10-3). Existing 3D maps or 3D models may also be manually supplemented with braille descriptions (figure 10-4) to make the products suitable for people with visual impairments.

Machine creation

Modern advancements bring exciting innovations. 3D printing, for instance, allows for intricate and durable tactile maps with varied textures to represent features, such as roads, buildings, and natural landscapes. Another approach incorporates thermographic printing (swell-forming), which raises specific areas of the paper when heated, creating cost-effective tactile maps. For enhanced accessibility, some maps integrate audio elements, such as touch-activated sensors connected to a voice guide, offering detailed explanations of each point of interest. Such approaches don't just make maps functional—they open a world of exploration, offering a multisensory experience that transforms how individuals with visual impairments perceive, study, explore, and navigate their environment.

Figure 10-3. An example of a map using a combination of clear varnish layering and digital printing. The layering of synthetic paints and varnishes is based on the original manual method of creating a tactile layer. Department of Geoinformatics, Palacký University Olomouc.

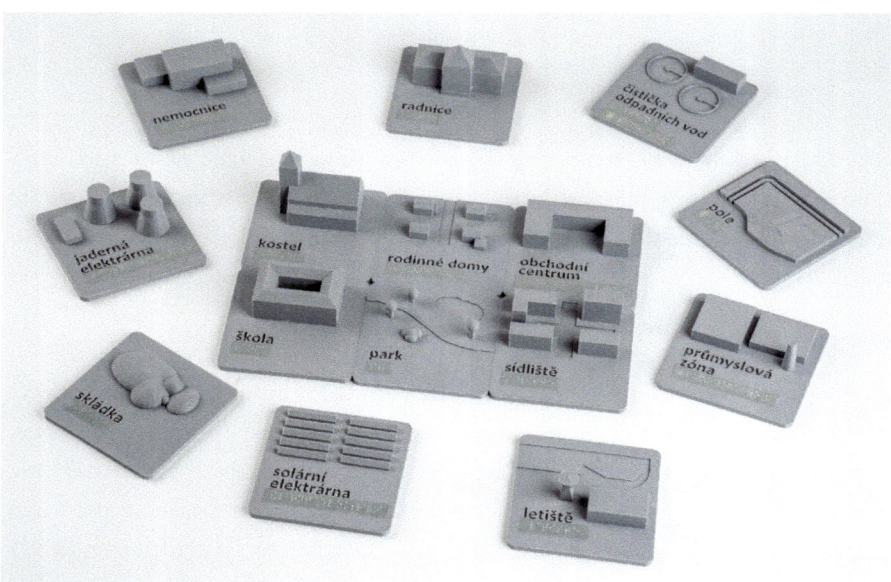

Figure 10-4. A teaching tool for understanding the spatial composition of a city, where individual "cards" with 3D objects are supplemented with descriptions in Latin script (printed in negative relief, making them nontactile) and manually labeled using a standard label maker to create braille captions. Department of Geoinformatics, Palacký University Olomouc.

Choosing the appropriate technique

Various methods are used for creating tactile maps, depending on the optimal workflow for the desired outcome. Some simple maps may be designed without an initial digital representation—for instance, drawing a simple plan using a black pen on microcapsule paper, which is then run through the swell-form machine to make a simple relief plan. However, most of the more advanced tactile maps need digital representation as the initial step before the manufacturing process. Such digital representations may be classified into two categories: 2D graphics and 3D models, eventually a combination of both. GIS plays a key role in providing source data and performing geospatial analyses to achieve topographic and thematic layers representing individual topics for tactile maps.

When one level is enough

Swell-forming is one of the most popular methods for creating tactile maps. This method is simple to design and cost-efficient. In this case, the digital representation is not far from an ordinary map. It consists of 2D graphics where black color represents symbols to be extruded into height (point, line symbols, textural fill of areas, braille labels, and other map elements). Such a map layout may, therefore, be easily designed using GIS software, using black color only for the elements to be raised over the rest of the map. Other colors may be used to enhance contrast between individual classes of map symbols, making the maps suitable for high-contrast reading and people with residual sight. Of course, an appropriate map generalization needs to be performed (see chapter 6, "Generalization for Tactile Maps"). After the digital representation is complete, the map layout is printed on swell paper and run through the swell-form machine, resulting in the touch-readable map.

Braille printers are another versatile tool for creating tactile maps. They were originally designed to represent written text but can now produce simple graphics as well. Braille printers excel at creating basic diagrams or geometric layouts, where raised dots form shapes that stand out against the background; their graphic capabilities are limited to a single height level. They seamlessly integrate braille descriptions directly onto the map, making them highly effective for accessibility. Modern braille printers can print on one or both sides of the page and work with tract paper or specially treated materials, offering flexibility for a variety of simple tactile applications.

More levels, more fun

When more than two height levels are required in the tactile maps, a digital 3D model needs to be designed instead of two-dimensional graphics. This typically involves more effort and specific skills, combining GIS knowledge with 3D modeling in external software. However, designing a geometrically correct 3D model is the only way to achieve 3D-printed tactile maps.

3D printing is an additive manufacturing process that creates three-dimensional objects from digital designs. It works by laying down thin layers of material and then fusing these layers to create a physical object. The world of 3D printing is as diverse as the creations it produces, with a variety of technologies tailored to different needs (figure 10-5). This applies especially when it comes to the production of multicolor 3D models. From fused deposition modeling (FDM) to stereolithography (SLA), selective laser sintering (SLS), and PolyJet, each technology demands a unique approach to model preparation.

FDM and FFF (fused filament fabrication), representing the most widely used 3D printers, typically require a single-file 3D file (STL format is the most typically used). However, creating multicolor prints (figure 10-6) might mean splitting the

Figure 10-5. The 3D printing technique provides an almost unlimited choice of shapes.
Department of Geoinformatics, Palacký University Olomouc.

Figure 10-6. A multicolor map showing population density in Czechia using the FFF method. Department of Geoinformatics, Palacký University Olomouc.

model into separate files by color or using software that supports a full-color slicing process.

SLA and SLS technologies, by contrast, focus on monochromatic precision, excelling in structural integrity over aesthetics. PolyJet and multi-jet fusion (MJF) take the spotlight for multicolor printing, allowing seamless integration of color directly into the model. These methods often require OBJ or 3MF file formats, which support intricate details such as textures and color gradients, whereas STL does not. Understanding these differences ensures that your digital masterpiece comes to life in vibrant hues or with pinpoint accuracy.

After choosing the desired configuration of 3D printing equipment, the designer needs to set an appropriate workflow for processing geospatial data. It typically begins with GIS, where input layers of various geometries are converted into polygons forming a gapless coverage of the mapped area. The polygons are later exported into 3D modeling software, where layers are extruded into volume. Such a process could be performed even in GIS. However, tools for exporting compatible 3D geometries between GIS and 3D modeling software are still lacking. As described earlier, coloring might be performed in the 3D modeling software or later within the slicing process.

Molding techniques

Another approach involves forming sheets of paper, plastic, or metal into specific shapes (figure 10-7). This is commonly applied not only to create traditional tactile maps but also to terrain models. It is an incredibly versatile process with applications across various materials and methods. For plastic, thermoforming is a popular approach (figure 10-8), where a heated sheet is shaped into a mold, often using vacuum forming to pull the material tightly around the mold to capture precise details. Vacuum forming excels in creating raised terrain models or tactile maps with intricate contours and relief, as the vacuum pressure ensures the material conforms closely to the mold. Nonvacuum techniques, such as manual pressing or embossing, can be used for simpler shapes or when working with less flexible materials, such as metal or thicker paper.

Paper embossing is a common nonvacuum method, often combined with preprinted designs to add tactile details to visually rich maps. Similarly, thin sheets of metal can be shaped through stamping or pressing, creating significantly more durable tactile models (typically for exterior use) that can also serve as reusable molds for other materials. A key advantage of these methods is their compatibility with color printing: Plastic sheets can be preprinted with vivid designs before forming, allowing for the creation of tactile maps that combine both touchable features and visual information. Paper and metal models can also incorporate color through preprinted layers or postforming processes, such as painting or screen printing. Thermoforming and embossing, therefore, may need both 3D and 2D digital inputs for storing information about the shape and texture separately.

Multimedia haptic maps

Traditional static maps are invaluable for conveying spatial information, but they often fall short in offering interactivity and accessibility for diverse user needs. Multimedia maps address these limitations by incorporating features such as sound, vibrations, and dynamic visual elements, enriching user experiences and fostering inclusivity. Building on these innovations, tactile maps push accessibility further by transforming spatial data into tactile formats through cutting-edge techniques. Recent advancements in 3D printing, haptic technologies, and audio navigation are significantly enhancing the functionality and precision of multimedia tactile maps. For example, specialized tactile aids, such as tactile pads, employ movable pins to create dynamic tactile patterns, enabling users to explore braille or graphic shapes with their fingers. Meanwhile, advanced systems use AI to analyze and segment

Figure 10-7. Single-color relief map of an urban area using various textures and braille labels. Department of Geoinformatics, Palacký University Olomouc.

Figure 10-8. The colored plastic sheet formed into a terrain shape. Department of Geoinformatics, Palacký University Olomouc.

images, producing tactile graphics optimized for people with visual impairments. Other innovations replace individual braille descriptions with conductive surfaces that deliver audio or haptic feedback, further improving user engagement and understanding. These developments highlight the growing potential of tactile maps to offer personalized and accessible mapping solutions for a wide range of users.

Added value to traditional haptic maps

Traditional haptic maps use a third dimension to convey information about the attributes of displayed objects, often supplemented by braille. When created according to established principles of tactile map design, these maps can effectively fulfill their purpose. However, their primary limitation remains the restricted amount of information they can convey because of the perceptual constraints of people with visual impairments—for instance, the maps often exclude braille to avoid overwhelming their users. Also, updates to these maps require the creation of new versions.

Cartographers have sought to enhance tactile maps with multimedia elements to address these challenges. Examples include replacing individual braille descriptions with audio components or incorporating braille codes and abbreviations into the interactive map legend. Audio integration allows dynamic voice feedback, providing more detailed contextual information and an enriched user experience. Such maps can also support real-time updates, transforming how users engage with spatial data.

The development of multimedia tactile maps reflects more than 40 years of technological evolution. During this time, numerous attempts were made to deliver fully functional solutions. Although early approaches were innovative, they were often complex to build, not user-friendly, and required significant effort to operate. Despite these challenges, these pioneering efforts laid the foundation for modern solutions, while also underscoring the need for diverse approaches to address varying user requirements.

Categories of interactive tactile maps

Interactive tactile maps can be divided into digital interactive maps (DIMs), which rely solely on digital platforms without physical overlays, offering accessible and user-friendly solutions, and hybrid interactive maps (HIMs), which include a haptic overlay on a digital display, combining traditional tactile maps with digital interactivity. Of course, it's possible to combine the strengths of DIMs and HIMs into an integrated approach.

Digital interactive maps

The fundamental building block of digital interactive maps is the display, which serves as the map itself. These solutions eliminate the need for physical overlays, relying solely on the screens of devices, such as tablets, mobile phones, or computers. Haptic feedback conveys information through the user interface, enabling users to perceive map data via touch. A key advantage of this approach is that it removes the necessity of registering a physical layer before interacting with the map. This process is often complex and challenging for people with visual impairments. By simplifying this step, digital interactive maps make tactile mapping more accessible and user-friendly.

Integrated approaches

Different technologies or approaches can be used to convey tactile feedback. An example is Braille Dis 9000, which integrates a GIS database and a large matrix array with pins. Some of the integrated access solutions use braille to interact with the user. A prime example is the Hyperbraille technology. This technology combines a tactile display (a matrix of piezo actuators) and a desktop computer. A piezo actuator is a device that uses the piezoelectric effect to produce precise mechanical movement or deformation in response to an electrical signal. Applications using flashing elements have also been developed. Another exciting technology is Linespace, which is used on a device called Homefinder. This system was developed at the Hasso Platner Institute in Potsdam, Germany. Its principle of operation is simple. Linespace uses a specialized plotting arm or mechanism to draw tactile lines on a surface. These lines can be refreshed or updated dynamically, making the display reusable. The individual user inputs are further monitored by a camera, primarily tasked with monitoring the user's movements and moving the extruder. Once the task has been completed, a scraper can clean the surface. The system is also capable of registering voice commands. The drawbacks of all integrated approaches are the difficulty in updating content and the relatively limited mobility. An advanced tactile display designed to improve accessibility for people with visual impairments is called Dot Pad. It uses an array of movable pins to create dynamic tactile patterns, allowing users to explore braille text, tactile graphics, and spatial representations in real time. The device connects to digital platforms including smartphones or computers, providing haptic feedback for interactive experiences.

Hybrid interactive maps

The main element of hybrid interactive maps is a tactile superstructure placed on the touch screen. The basic principle is the registration of the user's inputs through the haptic superstructure to the digital display. The digital display registers the location and triggers the appropriate interaction. These approaches aim to provide the user with additional information stored as multimedia content. Hybrid interactive maps help reduce the tactile complexity of maps, thus addressing the issue of tactile content accumulation.

Efforts to place a haptic surface over a display can be found as early as the 1980s, with one of the first attempts being a system called NOMAD. At that time, this system already used a digital tablet with a traditional tactile map placed on top. Over the years, several other systems have adopted this audio-tactile HIM approach. The advantage of these solutions is using the traditional tactile map in combination with a digital display. On the other hand, the difficulty of creation and the need for specialized hardware can be considered a disadvantage. This combination can sometimes be slow, expensive, and cumbersome. Additionally, if the physical map is not securely attached, it may lead to inaccurate activation of the digital map or cause the entire system to malfunction.

The Talking Tactile Tablet was considered a groundbreaking technical solution. The system used a touch screen, allowing users to place tactile sheets of swell paper on the tablet. Through these sheets, users then receive the desired information. The system is limited by the precision of printing the produced sheets, so the user must train to work with it. Another system, called Touchplates, is based on overlays made of acrylic, a durable and lightweight plastic that is smooth yet can hold raised tactile patterns (such as braille or raised lines) for users to feel. The system can recognize the given overlays automatically. Other authors have also explored similar approaches.

The combination of placing a conductive 3D surface on the display of a portable device with a touch screen then provides entirely new possibilities. Applications based on interaction with a conductive object can add interactivity to commonly used objects, whereas 3D printers are most used for creating conductive objects today. LucentMaps uses the conductive property of materials. The overlays in the form of maps are made of translucent plastic material with a height of up to 1.9 mm, which transmits touch. The map is made from conductive plastic, which allows the system to detect where a user touches it. When a user places their finger on a location, the system recognizes the touch and plays an audio description of that place.

If the user rotates the map, the system detects the change and adjusts the information accordingly, ensuring that directions and spatial references remain accurate. This makes it easier for visually impaired users to understand locations and navigate spaces effectively.

TouchIt3D technology was developed in 2014 at Palacký University in Olomouc, Czechia. This technology allows for combining two or more materials through 3D printing, with at least one being conductive. This enables the creation of printed models that function as interactive tactile units. TouchIt3D is used to produce tactile maps that incorporate spatial orientation features along with haptic or audio feedback. The models include conductive elements linked to a software application that assigns specific actions to detected signals, such as triggering sound or vibration responses when touched. A key difference between TouchIt3D and LucentMaps lies in the design of the conductive parts, which serve as the interactive points in TouchIt3D. These models become interactive when placed on a capacitive display, enabling additional functionality. The primary application of TouchIt3D is the creation of interactive tactile maps (figure 10-9).

Figure 10-9. The tactile map of street layouts produced using TouchIt3D technology. Department of Geoinformatics, Palacký University Olomouc.

Challenges and future of multimedia tactile maps

Despite their innovative features, multimedia tactile maps have limitations. They are often not weather-resistant, making them unsuitable for outdoor use. Many rely on electronic components that need regular charging or power supply, which can be inconvenient in remote or extended-use scenarios. Some systems are bulky or fragile with limited portability. Their production costs can also be high, mainly when using advanced haptic systems. Furthermore, creating and using these maps often requires specialized skills, which may limit their accessibility for broader audiences or less experienced users.

However, with ongoing technological advancements, we are witnessing a new era of multimedia maps, significantly shaped by AI, that introduces possibilities such as reading and interpreting screen content, generating tactile patterns or images from textual information, and recognizing objects, places, or routes in real time. These capabilities enhance the interactivity and functionality of tactile maps, making them more adaptable to diverse user needs. In addition, developing powerful processors, increasingly portable mobile devices, and improved intellectual property protection offers a promising future for multimedia tactile maps. These advancements enable the creation of robust and weather-resistant solutions that are easier to use and maintain, paving the way for broader accessibility and application in various environments. As technology continues to evolve, the potential of multimedia tactile maps to transform accessibility and user experience grows exponentially.

Conclusion

Multimedia tactile maps represent a transformative leap in accessibility, combining haptic feedback, audio elements, and advanced digital technologies to create a rich, interactive experience for users with visual impairment. Although traditional tactile maps rely on braille and static tactile designs, the integration of multimedia elements addresses their limitations by enabling dynamic interactions and real-time updates. Such innovations use tools, such as 3D printing, audio navigation, and AI-driven tactile pattern generation, to expand the functionality of maps and enhance user engagement.

Notably, not all applications require such high-tech solutions. In some scenarios, manually crafted tactile maps may be entirely sufficient. For others, 3D-printed models offer a cost-effective and precise alternative. Meanwhile, the potential of multimedia maps shines in environments demanding real-time feedback and interactive exploration. Could this range of approaches inspire new ways to meet the diverse needs of people with visual impairments? What challenges remain in making these technologies universally accessible? By exploring the possibilities across manual, 3D-printed, and multimedia solutions, the future of tactile mapping invites continued innovation. Finally, we continue to seek ways to refine these tools to create more meaningful and accessible spatial experiences for everyone.

Further reading

In the archives of the Perkins School for the Blind, there is a preserved tactile map of South America from approximately 1900. See www.flickr.com/photos/perkinsarchive/26055187663/in/album-72157658605773326.

A detailed listing of the different approaches of multimedia maps can be found in Sile O'Modhrain, Nicholas A. Giudice, John A. Gardner, and Gordon E. Legge, "Designing Media for Visually-Impaired Users of Refreshable Touch Displays: Possibilities and Pitfalls," *IEEE Transactions on Haptics* 8, no. 3 (July 1, 2015): 248–57.

On specifics on DIM

Applications using flashing elements: B. Schmitz and T. Ertl, "Interactively Displaying Maps on a Tactile Graphics Display," in *SKALID 2012–Spatial Knowledge Acquisition with Limited Information Displays*, 2012, 13–18.

Braille Dis 9000: Thorsten Völkel, Gerhard Weber, and Ulrich Baumann, "Tactile Graphics Revised: The Novel Braille Dis 9000 Pin-Matrix Device with Multitouch Input," in *Proc. 11th Int Conf Computers Helping People with Special Needs*, Linz, Austria (Berlin: Springer, 2008), 835–42.

Hyperbraille technology: Thomas Kieninger and Norbert Kuhn, "Hyperbraille: A Hypertext System for the Blind," in *Proceedings of the first annual ACM conference on Assistive technologies*, 1994.

Integrated approaches: Timo Götzelmann, "Visually Augmented Audio-Tactile Graphics for Visually Impaired People," *ACM Transactions on Accessible Computing (TACCESS)* 11 (2): 1–31.

Linespace: Saiganesh Swaminathan, Thijs Roumen, Robert Kovacs, David Stangl, Stefanie Mueller, and Patrick Baudisch, "Linespace: A Sensemaking Platform for the Blind," in *Proceedings of the 2016 CHI Conference on Human Factors in Computing Systems*, 2016.

On audio-tactile HIM approaches

Haptic Soundscapes: Daniel Jacobson, "Haptic Soundscapes: Developing Novel Multi-Sensory Tools to Promote Access to Geographic Information," in *WorldMinds: Geographical Perspectives on 100 Problems*, ed. Donald G Janelle, Barney Warf, and Kathy Hansen (Springer, 2004), 99–103.

LucentMaps: Timo Götzelmann, "LucentMaps: 3D Printed Audiovisual Tactile Maps for Blind and Visually Impaired People," in *Proceedings of the 18th International ACM SIGACCESS Conference on Computers and Accessibility (ASSETS '16)*, Association for Computing Machinery, New York, 2016, 81–90.

Mobile Touch Screens: Shaun K. Kane, Jeffrey P. Bigham, and Jacob O. Wobbrock, "Slide Rule: Making Mobile Touch Screens Accessible to Blind People Using Multi-Touch Interaction Techniques," in *Proceedings of the 10th International ACM SIGACCESS Conference on Computers and Accessibility (ASSETS '13)*, Association for Computing Machinery, 2013, 73–80.

Multimodal: Bruce J. Holmes et al., "Multimodal Interfaces in Geographical Information Systems," *Journal of Visual Impairment & Blindness* 90, no. 5 (1996): 389–93.

NOMAD: D. Parkes, "NOMAD: An Audio-Tactile Tool for the Acquisition, Use, and Management of Spatially Distributed Information by Partially Sighted and Blind People," Second International Conference on Maps and Graphics for Visually Disabled People, Nottingham, 1988.

Talking Tactile Tablet: Stanley Landau and Lyle Wells, "Merging Tactile Sensory Input and Audio Data by Means of the Talking Tactile Tablet," in *Proceedings of EuroHaptics* 3, Association for Computing Machinery, 2003, 414–18.

TouchIt3D: Jan Brus, Radek Barvíř, and Andrea Vondráková, "Interactive 3D Printed Haptic Maps—TouchIt3D," in *Joint ICA Workshop Cartography for Specific Users*, ed. Carla Cristina Reinaldo Gimenes de Sena, Barbara Flaire Jordão, and José Jesús Reyes Nuñez, International Cartographic Association, Tokyo, 2019. Similar concepts, which involve interactive 3D models rather than tactile maps, can be found in Michael A Kolitsky, "3D Printing Makes Virtual World More Real for Blind Learners," *E-Mentor* 1, no. 63 (2016): 65–70.

Touchplates: Shaun K. Kane et al., "Touchplates: Low-Cost Tactile Overlays for Visually Impaired Touch Screen Users," in *Proceedings of the 2013 ACM SIGCHI Conference on Human Factors in Computing Systems (CHI '13)*, Association for Computing Machinery, 2013, 935–44. Similar solutions are proposed by André M. Brock, *Interactive Maps for Visually Impaired People: Design, Usability and*

Spatial Cognition 3, Université Toulouse, Paul Sabatier, 2013; André M. Brock, "Touch the Map! Designing Interactive Maps for Visually Impaired People," *ACM SIGACCESS Accessibility and Computing*, no. 105 (2013): 9–14; André M. Brock et al., "Interactivity Improves Usability of Geographic Maps for Visually Impaired People," *Human–Computer Interaction* 30, no. 2 (2015): 156–94; E. Brulé et al., "MapSense: Multi-Sensory Interactive Maps for Children Living with Visual Impairments," in *Proceedings of the 2016 CHI Conference on Human Factors in Computing Systems*, ACM, 2016, 445–57; David McGookin, Stephen Brewster, and Weiwei Jiang, "Investigating Touchscreen Accessibility for People with Visual Impairments," in *Proceedings of the Fifth Nordic Conference on Human-Computer Interaction* (*NordiCHI '08*), Association for Computing Machinery, New York, 2008, 298–307; Chloe Senette, Stéphane Guillaume, and Thibaut Bourdeaud'Huy, "Tactile Maps for the Visually Impaired: Principles of Design and Evaluation," *Journal of Visual Impairment & Blindness* 107, no. 3 (2013): 231–42.

For an in-depth exploration of tactile cartography, including advanced techniques and detailed case studies, go to https://tactilemaps.upol.cz.

About the authors

The authors are based at the Department of Geoinformatics, Faculty of Science, Palacký University Olomouc, Czechia. They have all contributed to numerous projects focused on creating tactile maps and 3D models for individuals with visual impairments. They are coauthors of two monographs on tactile maps and the developers of the TouchIt3D technology for interactive 3D-printed tactile maps.

Radek Barvíř works as an assistant professor of advanced visualization. His research interest is enriching map production efficiency, thematic cartography, and geospatial 3D printing. He is the author of *Graphic Map Load Measuring Tool for Assuming the Visual Complexity of Maps* and one of the developers of the online Value-Scale Generator.

Radek Barvíř.

Alena Vondráková is a passionate cartographer and graphic designer, working as an assistant professor. She leads cartography courses and focuses her research on thematic maps, atlases, and user issues. She is the author or coauthor of more than 10 cartographic books and atlases. Tactile cartography has been her passion for more than 15 years.

Alena Vondráková.

Jan Brus is an assistant professor specializing in environmental geoinformatics and innovative mapping solutions. His research combines spatial data with advanced technologies, mainly 3D printing. Jan's work bridges the fields of geoinformatics, accessibility, and technology with applications ranging from navigation aids and educational tools to solutions for visually impaired users. He remains dedicated to exploring new possibilities in tactile mapping and multimedia interactive solutions.

Jan Brus.

Chapter 11

Methodical reflections

Albina Mościcka

One day, a student came to me. He was interested in 3D printing and wanted to use it in cartography. Our ideas soon turned toward tactile maps. We didn't know much about the topic, but it quickly became apparent that the new student was very inquisitive and would be a good candidate for a PhD student. And thus began my journey into the world of methodical tactile mapping.

Before he started his doctoral studies, we met with one of the most important Polish tactile cartographers of the time. We wanted to begin collaborating and drawing on his experience. However, the results were not good. We were a little discouraged by his words, that 3D printing would not work in such studies because blind users do not like new solutions. However, we had enough courage to say that scientists are not only there to fill in the gaps in existing knowledge but are also obliged to create new solutions and convince recipients to use them. We firmly believed that if we did it our way, we would gain the trust of users.

Today, after seven years of experience, Jakub's doctoral thesis defended with distinction, one research project successfully completed and another in progress, publications in prominent scientific journals, and, most importantly, recognition from users of our tactile maps, I can say with full conviction that we were right. And the key to success was a methodical approach to tactile mapping based on the involvement of blind users at every stage of the process.

—Albina

Introduction

Creating maps for people with visual impairments is a challenging task, and not only because of their different modes of space cognition and other requirements for the map content presentation. One of the main difficulties is our inability to quickly assess whether the solutions proposed by the cartographer are legible and understandable for the recipient, especially because there are no standards or specific guidelines on how such maps should be edited. In traditional cartography, the first assessment and the correction of the map made can be performed by its author. Looking at our own map, we see that something is illegible and requires additional editing or correction. We can also ask colleagues to check the compliance with technical requirements or to find errors that we didn't notice.

We do not have such options when developing tactile maps. Usually, there are no blind people around us who could immediately verify the effect of our work. What's more, traditional printing methods used so far have been expensive, so tactile cartographers haven't had access to cheap and fast printing of their maps. All this means that tactile maps have often left much to be desired in terms of quality and were not always accepted by users.

The development of additive techniques, including 3D printing, has changed the situation in tactile cartography. Printing techniques have emerged that allow rapid prototyping—that is, printing a tactile map prototype quickly and cheaply, correcting and verifying it. As a result, there has been a recent renaissance of interest in the development of tactile maps. However, easy access to 3D printers and software may lead one to believe that anyone can create a tactile map today. In theory, yes. But can anyone make a good tactile map? What exactly is a good tactile map?

A good tactile map is a map that is legible by people with visual impairments, safe and comfortable for the user, and understandable in terms of information transfer. Legibility is the ability to recognize and distinguish between signs and their meanings. The less effort that users must put into this activity, the better, because then they can concentrate on the information about the spatial relationships conveyed by the map, not only on decoding the meaning of individual signs. The comfort of using a map is the safety and convenience of using it. Informativeness refers to the amount of knowledge about spatial relationships between objects or phenomena that users gain from the content of the map. The goal of any map is to ensure that this knowledge—the user's idea of space—is as close to reality as possible.

The development of a map that meets all these requirements requires extensive expert knowledge not only in the field of cartography, but also in information theory, tactile perception, special pedagogy, and so on. Therefore, tactile mapping requires cooperating with specialists from other fields and disciplines and, above all, openness to the unique needs and perceptual abilities of the recipient—i.e., the user and reviewer of the map. In addition, maps should always be produced according to the same principles, and the effectiveness of these rules should be verified in practice by the recipients. This will ensure repeatability in the production of high-quality maps.

In this chapter, I will present selected aspects of the methodical approach to tactile mapping. It is based on the experience gained during the last seven years of work on this topic, and especially during the three-year project on the technology of developing tactile maps of historical gardens,[1] which was completed in July 2024. My reflections are based on two fundamental assumptions. The first is the participatory and inclusive approach. Participation implies the involvement of users at all stages of tactile mapping. Each solution proposed by the cartographer should be evaluated by a group of people with visual impairments. Their suggestions are the basis for changes, corrections, and additions. Once introduced, they are reevaluated until they are accepted by the target user group. Inclusivity, on the other hand, is related to the development of solutions for a wide group of recipients—people with different characteristics, different disabilities, and different abilities. The group of people with visual impairments evaluating the cartographer's work should therefore also be differentiated accordingly.

The second assumption is that user evaluation is conducted in a methodical, systematic manner. This implies the adoption of strictly defined principles, techniques, and procedures aimed at avoiding randomness of results and obtaining information about the correctness of the evaluated solutions. These principles can be based on the methods and rules used in the development of services and products for traditional recipients, only skillfully adapted to the specific perception of users with special needs. Considering that the most important aspects of tactile maps are their legibility, ease of use, and accuracy of information transfer, user evaluation should verify whether each of these criteria has been met. A methodical approach creates the conditions for ensuring high reliability and repeatability of the obtained results.

Testers

Including people with visual impairments in all phases of tactile mapping is not easy for many reasons. First, this group is diverse in terms of visual impairment and tactile ability. There are people who lost their sight in childhood or as adults, but there are also those who have never been able to see. There are people who are completely blind and those who are severely visually impaired but have the possibility of residual vision—to very different degrees. In addition, this group includes people who are very well rehabilitated, with extensive experience in working with tactile materials (and maps) and knowledge of the braille alphabet. But there are also those whose tactile skills and experience are negligible.

Therefore, when developing tactile maps for such a group of recipients, their diversity should be considered, and the evaluation of the proposed solutions should also be carried out by a group of testers with equally diverse characteristics. Their sociodemographic characteristics, such as age, gender, and education, should also be considered, as these may have an impact on the reception and understanding of tactile maps. The diversity of the group is particularly important when developing maps for a wide range of users, such as tourist tactile maps. This poses an additional challenge when choosing the composition of the test group, which should be as diverse as the group of end users of tactile maps and big enough to ensure adequate representativeness of the study group.

Organizing a large, diverse test group is also complicated by the fact that this is a dispersed group of recipients, and because tactile materials are meant to be touched and seen, these people must be physically present at the study sessions. So, it is extremely difficult to gather several dozen people with different characteristics in one place at the same time, especially when mobility is a big challenge for many of them.

Such difficulties are probably the reason that the work done so far has not been verified on methodically selected groups of people with visual impairments and analyzed statistically. Because of the limitations in selecting a diverse group, the optimal sample size cannot be precisely defined.

Methods

In this chapter, I am going to share my reflections on three basic aspects of methodical tactile mapping: (1) the legibility of tactile signs and the principles of map editing, (2) the comfort and profitability of the reproduction methods used, and (3) the usefulness and clarity of the information given by tactile maps. These solutions are based on the principles of traditional cartography but also on methods used in the social sciences (tactile pedagogy) and information systems theory. They have been adapted to the needs of tactile perception and have been used in research experiments for the development of tactile maps of historical gardens, with very good results and consequently user satisfaction. They can therefore be used to evaluate other tactile maps.

Tactile signs and editing rules

Tactile signs used on maps are still cartographic signs, so they are subject to the same rules as classic cartographic signs. Thus, the theory of cartographic sign construction, which treats the sign as a semiotic object, should also be applied to tactile cartographic signs. The laws of semiotics concerning communication through sign systems are applied in map language. They include three types of relations: semantic (between signs and reality), syntactic (between signs), and pragmatic (between signs and map recipients).[2] Therefore, the methodical evaluation of tactile signs should include the verification of the correctness of these semiotic relations, which can be applied in a two-step method.

The first step of the method involves checking the correctness of semantic and syntactic relations. At this stage, only syntactic relations between signs of the same geometry are verified. In practice, this is realized with the help of stimulus matrices prepared separately for point, line, and area signs. This method is called "in isolation" verification. The matrices are based on a proposed set of tactile signs. The signs on the matrices are randomly arranged in rows and columns and are duplicated to make the task more difficult (figure 11-1). Each matrix has a legend explaining all the signs shown on the matrix.

The person with visual impairments is asked to find the signs with the given meanings in the rows of the matrix. Failure to assign a sign to a given meaning or failure to find a sign within a given time is an error. The task examines the ease of decoding signs and assigning them to meanings (semantic relations) and the differentiation of signs within the same geometric categories (syntactic relations). The results show which signs are legible and easy to find, which are not, and which signs are confused with one another. Signs that cause errors must be corrected and retested. When all signs are considered legible and distinguishable by people with visual impairments, the second stage of the process can begin.

In the second step, the correctness of the semantic, syntactic (between signs of all types of geometry), and pragmatic relations is verified. At this stage, the influence of other signs, including those with different geometries, is evaluated. In practice, signs with different geometries are placed on maps, so this procedure is called "in context" verification. Pseudo maps can be used in this step. They are maps that imitate but do not represent reality and are used to place signs in many combinations on one sheet (figure 11-2). The legend explaining all the signs is also developed.

Two types of tasks are proposed to examine signs in context. In the first task, people with visual impairments are asked to find the signs on the map and, in the

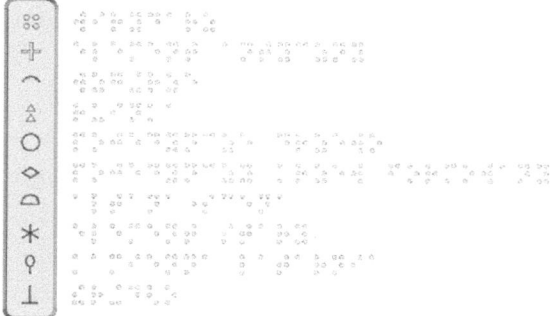

Figure 11-1. Matrix for point signs with legend, 2D version. Source: Materials of the Technology of Developing Tactile Maps of Historical Gardens project.

Figure 11-2. Pseudo map of a Japanese garden, 2D version. Source: Materials of the Technology of Developing Tactile Maps of Historical Gardens project.

second, to tell the story of what is in the map. The process requires finding and distinguishing signs, and then assigning meaning and interpreting them, which complements the effectiveness of the first task.

The results of the contextual sign verification allow us not only to identify the illegible signs among signs of different geometry but also, at this stage, to verify the rules of map editing, such as distances between signs. This is because signs can be classified as illegible not because of their geometric parameters but because they're impossible to distinguish from the environment—for instance, a point sign placed on an area sign can merge with its pattern. Therefore, the separation of errors resulting from geometric parameters of signs and from editing rules is a crucial challenge at this stage of the verification. Regardless of the source of the errors, only after full acceptance of the signs and editing rules by people with visual impairments can all semiotic relations be considered correct.

Printing

The quality of a tactile map is also influenced by the printing technique. Widely used relief printing techniques, such as thermoforming, thermal printing, swell paper, and manual methods, often fail to meet the cartographers' expectations.[3] That's why we started looking for new printing techniques. In recent years, methods that build physical objects layer by layer based on computer models have become increasingly popular.[4] They enable us to print the same map model that was designed and to have control over map parameters. A well-designed model is one that allows the operator to control the printing parameters and that is legible. The printing technique also affects the user's comfort in working with the printed map. Practical aspects are also important: the cost of printing a map, the time needed to prepare the map for printing, or resistance to weather conditions. Therefore, the tactile map printing technique should be verified by considering two main aspects: user experience and production efficiency.

In tactile pedagogy, the usefulness of various tactile materials can be evaluated, for example, through user experience in three areas: practical use, semantic differential, and ranking methods.

The practical use verifies map legibility and is based on a person with visual impairments' assessment of their ability to perform location tasks on different substrates, examining the influence of the printing technique on task completion time and accuracy (figure 11-3). Tasks should be associated with all geometric sign types (point, line, and area), and the number of errors should be documented.

The semantic differential uses a series of bipolar adjectives or phrases to evaluate the user's perception and evaluation of a concept or object. Respondents rate adjectives on a five-point Likert scale, which assesses the emotional and cognitive perceptions across multiple dimensions.[5] In the case of tactile maps, these are haptic (for instance, rating from soft to hard), functional (for instance, rating from uncomfortable to comfortable), and emotional (for instance, rating from "I like it" to "I don't like it") dimensions. To reduce bias, the order of terms varies, with positive ratings associated with both lower and higher numbers.[6]

The final stage of the user experience evaluation is the ranking of the maps. Participants are asked to identify the best and worst maps—for instance, three in each group.

The second aspect of evaluating the printing technology focuses on practical and economic concerns. First, there is the question of cost—the price of printing one sheet of a map. The lower it is, the better. However, the cost of printing one

Figure 11-3. A user testing tactile map samples that were printed using various printing techniques. Image courtesy of Jakub Wabiński.

sheet is influenced by the size of the print run and the time (difficulty) required to prepare the map for printing. Thermoforming and thermal printing produce maps that are legible and pleasant to the touch and enable simultaneous printing of tactile and color elements. However, they make it possible to achieve a low unit price only with large print runs, because they require the preparation of an expensive matrix. Some additive techniques (SLA, UV printing) make it possible to print a single map at an attractive price, but preparing a map for printing and operating the printer requires advanced skills and a longer printing time. Other 3D printing techniques yield comfortable tactile maps, but they are very expensive (PolyJet). The cheapest and simplest is still printing on swell paper. However, this does not allow for control over map parameters and provides a fragile paper map.

Today, it is extremely important to consider universal design when creating products for people with special needs. In this context, the development of hybrid maps—maps with both tactile and visual content—has become a standard. They contain convex content and a colored base. Both should be printed at the same time. Such technologies are also available among modern printing technologies (PolyJet—3D printing and UV printing).

In addition, the evaluation should consider how the maps will be used (indoors, outdoors) and the effect of weather conditions on their durability. Because these practical evaluations are so individual, there are no procedures in the literature for evaluating printing techniques. They can therefore be adapted to one's own needs.

In summary, the evaluation of the suitability of a printing technique should be considered from two angles: the evaluation of the user experience and the evaluation of the needs of the authors. The first part of the evaluation can be treated as constant, independent of the authors' expectations, whereas the second part is variable. The authors' expectations may depend on their budget or skills. For example, for some, the lowest price may be the most important factor, whereas others may prioritize the highest quality product, regardless of cost. These considerations can also change over time, due to the decreasing prices of modern printing equipment and services, as well as the increasing sophistication of authors.

Usefulness and ease of use

Because tactile maps serve a specific purpose, all new cartographic works should be verified by users in terms of legibility and information content (figure 11-4). Methods used in information technology can be adapted for this purpose. Recently, this field has been faced with the need to verify whether newly created applications are easy to use and useful for users. One of the methods for such verification is the technology acceptance model (TAM).[7] TAM focuses on two theoretical constructs: perceived usefulness and perceived ease of use, which are fundamental determinants of information system usage.

Adapting terms from IT to tactile mapping, perceived usefulness can be defined as "the degree to which a person believes that using a particular tactile map will bring them some benefit"—for instance, obtaining new information that is not available from other sources or in other ways. One such type of information when exploring an area is the spatial relationships between objects. People with visual impairments can neither see nor touch them. A tactile map, reflecting the real relations between objects located in space, allows them to fully explore areas by touching the reduced image of this part of the Earth.

Perceived ease of use refers to "the degree to which a person believes that using a particular map would be effortless" and can be defined in the same way for both systems and maps. In the case of maps, ease of use is strongly related to the legibility and comfort of the map. Legibility is the most important aspect of all tactile maps. The more legible a map, the easier it is to use and the more likely it is to be accepted by users. At the same time, although the elements on the map must be legible, they

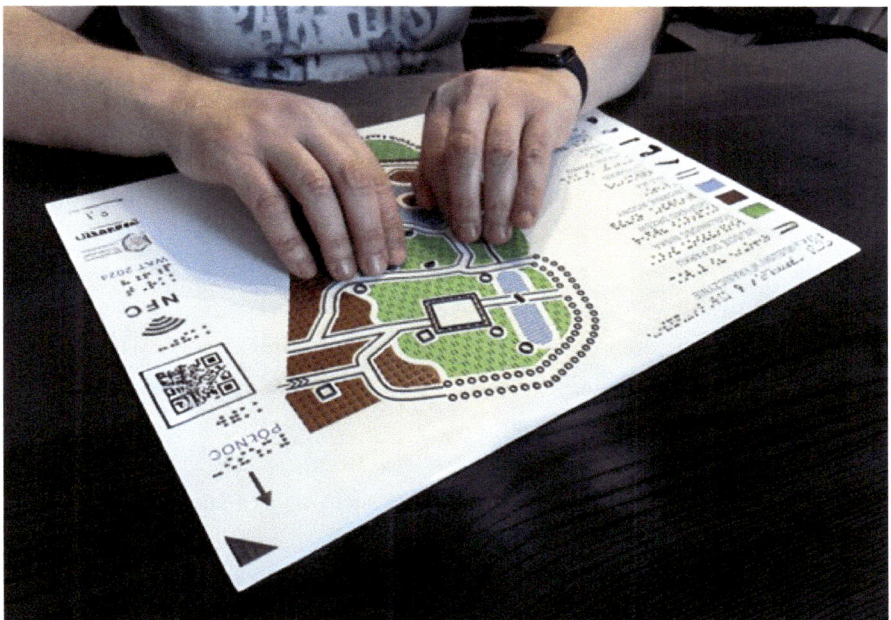

Figure 11-4. A user testing the legibility and information content of a tactile map of a garden. Image courtesy of Jakub Wabiński.

cannot be too sharp, so as not to hurt the user's fingers. Thus, the safety of touching and the comfort of using the map are also part of its ease of use.

In practice, the implementation of TAM is achieved through the appropriate composition of the research test.[8] From the definitions of perceived usefulness and perceived ease of use, initial items are generated—statements that match the content of our determinants. As many of such statements as possible should be collected.

The initial elements of perceived usefulness might take the form of the following statements:

- Without a tactile map, it would be harder for me to learn about…
- The tactile map saves me time.
- The tactile map allows me to gain knowledge that I can't get from other sources.

The initial elements of perceived ease of use may take the form of the following statements:

- The signs on the map are easy to distinguish.
- The map content is easy to understand.
- The map is pleasant to the touch.

Based on the collected items, a preliminary test is performed. It includes two tasks: ranking and clustering. Ranking means arranging items from the most important to the least important, whereas clustering aims to collect items into groups of items with similar meaning. Both tasks aim to reduce the number of final items in each construct to a maximum of 10. The final items are then scored by people with visual impairments on a 5-point Likert scale.[9]

Both factors determine the user's attitude toward using a map, which is often included as an additional construct in TAM, called intention to use.[10] It may be defined as "the extent to which a person intends to use a particular map" and assessed using a simple choice between "I plan to use the map" or "I do not plan to use the map."

The credibility of the obtained test results depends on the level of reliability of the test. The reliability can be estimated by calculating the mutual consistency between scores for the test questions. It can be indicated by the coefficient Cronbach's alpha.[11] This indicator shows whether all items in the test measure the same concept. The value of the coefficient is always between 0 and 1. The closer the value is to 1, the more reliable the test. A score of 0.8 or higher is considered good to very good.

Conclusion

High-quality maps are primarily those that can be created in a controlled and repeatable manner. This demands a methodical approach, in which maps are created according to the adopted methods and rules. Such an approach minimizes the subjectivity of the map development process. The methodical verification of map correctness provides measurable results, contributes to objective evaluation, and improves the credibility of the results. All of this allows us to develop better maps that are more readily used by their recipients.

Developing maps for people with special needs requires a tailored approach that actively involves users throughout the process. This engagement helps us better understand their needs, resulting in a product that is user-tested, refined, and fully adapted to their requirements. Involving people with visual impairments in tactile mapping is also a great lesson in humility. No one likes to be judged. It is especially painful when that assessment concerns the results of many months of work, overturns many of our assumptions, and imposes new obligations. However, it is worth remembering that if we methodically check the work of the tactile cartographer at every stage of the map's development, the result may pleasantly surprise us. A clear and legible map is eagerly used by people with visual impairments, including those who have given up using tactile maps because of their insufficient quality. In the end, the gratitude and appreciation of the users is the best form of satisfaction (figure 11-5).

Figure 11-5. Students from the Laski Educational and Training Centre for Blind Children explore Romantic Park in Arkadia using a tactile map. Image courtesy of Jakub Wabiński.

Further reading

On sample selection and analysis of dependencies between test results and tester characteristics

Dunn, Olive Jean. "Multiple Comparisons Using Rank Sums." *Technometrics* 6, no. 3 (1964): 241–52.

Faulkner, Laura. "Beyond the Five-User Assumption: Benefits of Increased Sample Sizes in Usability Testing." *Behavior Research Methods Instruments & Computers* 35, no. 3 (2003): 379–83.

King, Bruce M., Patrick J. Rosopa, and Edward W. Minium. *Statistical Reasoning in Psychology and Education.* Wiley, 2003, 496.

Kruskal, William H., and W. Allen Wallis. "Use of Ranks in One-Criterion Variance Analysis." *Journal of the American Statistical Association* 47, no. 260 (1952): 583–621.

Mann, Henry B., and David R, Whitney. "On a Test of Whether One of Two Random Variables Is Stochastically Larger than the Other." *Annals of Mathematical Statistics* 18, no. 1 (1947): 50–60.

Ungar, Simon, Mark Blades, Christopher Spencer, and Kathryn Morsley. "Can Visually Impaired Children Use Tactile Maps to Estimate Directions?" *Journal of Visual Impairment & Blindness* 88, no. 3 (1994): 221–33.

US Department of Health and Human Services. Food and Drug Administration. *Applying Human Factors and Usability Engineering to Medical Devices: Guidance for Industry and Food and Drug Administration Staff.* 2016. www.fda.gov/regulatory-information/search-fda-guidance-documents/applying-human-factors-and-usability-engineering-medical-devices.

On tactile signs development and testing

Brittell, Megen E., Amy K. Lobben, and Megan M. Lawrence. "Usability Evaluation of Tactile Map Symbols Across Three Production Technologies." *Journal of Visual Impairment & Blindness* 112, no. 6 (2018): 745–58.

Jehoel, Sandra, Simon Ungar, Don McCallum, and Jonathan Rowell. "An Evaluation of Substrates for Tactile Maps and Diagrams: Scanning Speed and Users' Preferences." *Journal of Visual Impairment & Blindness* 99, no. 2 (2005): 85–95.

Mościcka, Albina, Emilia Śmiechowska-Petrovskij, Jakub Wabiński, Andrzej Araszkiewicz, and Damian Kiliszek. "Methodical Testing of Tactile Cartographic Signs in Isolation and in Context." *Cartography and Geographic Information Science* (2024): 1–18.

Perdue, Nicholas A., and Amy K. Lobben. "Understanding Spatial Pattern Cognition from Tactile Maps and Graphics." *Cartographica: The International Journal for Geographic Information & Geovisualization* 51, no. 2 (2016): 103–10.

On 3D printed tactile maps

Gual-Ortí, Jaume, Marina Puyuelo-Cazorla, and Joaquim Lloveras-Macia. "Improving Tactile Map Usability Through 3D Printing Techniques: An Experiment with New Tactile Symbols." *The Cartographic Journal* 52, no. 1 (2013): 51–57.

Kocovic. Petar. *3D Printing and Its Impact on the Production of Fully Functional Components: Emerging Research and Opportunities.* Hershey: IGI Global, 2017, 164.

Rener, Roman. "The 3D Printing of Tactile Maps for Persons with Visual Impairment." In *Universal Access in Human–Computer Interaction. Designing Novel Interactions: Lecture Notes in Computer Science*, edited by Margherita Antona and Constantine Stephanidis. Springer, 2017.

On tactile maps' usefulness

De Oliveira, Sabrina, Katsuk Suemitsu, and Maria Lucia Okimoto. "Design of a Tactile Map: An Assistive Product for the Visually Impaired." In *Advances in Ergonomics in Design Advances in Intelligent Systems and Computing*, edited by Fernando Rebelo and Maria João Soares. Springer, 2016.

Espinosa, Mangeles A., and Esperanza Ochaíta. "Using Tactile Maps to Improve the Practical Spatial Knowledge of Adults Who Are Blind." *Journal of Visual Impairment & Blindness* 92, no. 5 (1998): 338–45.

On the technology acceptance model

Van de Veen, Evelyn. Materials from the Written Examinations course. 2014.

Venkatesh, Viswanath, and Fred D. Davis. "A Theoretical Extension of the Technology Acceptance Model: Four Longitudinal Field Studies." *Management Science* 46, no. 2 (2000):186–204.

Acknowledgments

The importance of a methodical approach to tactile mapping was fully confirmed during the implementation of the apparently crazy idea of showing historical gardens to people with visual impairments using tactile maps. I would like to thank Jakub Wabiński, Andrzej Araszkiewicz, and Damian Kiliszek for this adventure. We could not have done it without the expert support of the tactile pedagogue Emilia Śmiechowska-Petrovskij from the Cardinal Stefan Wyszyński University in Warsaw and the specialist in garden art—Anna Traut-Seliga from the Stefan Batory Academy of Applied Sciences in Skierniewice. The Polish Association of the Blind also made an invaluable contribution by organizing eight test sessions and a series of consultations with people with visual impairments. A special thanks to our testers. It wasn't always easy to accept their comments, but they're what make these maps so good. I still can't believe we did it.

About the author

Albina Mościcka is associate professor, Faculty of Civil Engineering and Geodesy, Military University of Technology (MUT), Warsaw, Poland. She holds an MSc in geodesy and cartography and a PhD and Habilitation in Technical Sciences in the discipline of geodesy and cartography. She gained professional experience at Stanford University (USA), the University of Stuttgart (Germany), and as a key expert in cartography in EU-funded projects in Dushanbe (Tajikistan). At MUT, Albina has worked on the use of geoinformation technologies in the humanities, mainly in the field of management, access, and popularization of movable cultural heritage, as well as on tactile mapping. When not working on maps, Albina is interested in politics, cooking, and Georgian wines.

Albina Mościcka.

Notes

1   See case study "Tactile Maps of Historic Gardens" in part 3.
2   Atkin, Albert, "Peirce's theory of signs," *Mind* 119, no. 475 (2010): 852–55.
3   Buchroithner, Manfred F., and Claudia Knust, "True-3D in Cartography—Current Hard- and Softcopy Developments," in *Geospatial Visualisation: Lecture Notes in Geoinformation and Cartography*, ed. Antoni Moore and Igor Drecki (Berlin, Heidelberg: Springer, 2013), 41-65.
4   Srivatsan, Tirupathi S., and Tiruvengadam S. Sudarshan, *Additive Manufacturing: Innovations, Advances, and Applications* (CRC Press, 2015), 460.
5   Ploder, Andrea, and Anja Eder, "Semantic Differential," *International Encyclopedia of the Social & Behavioral Sciences*, 2nd ed. (2015): 563–71.
6   Babbie, Earl, *The Basics of Social Research* (Cengage Learning, 2016), 560.
7   Davis, Fred D., *A Technology Acceptance Model for Empirically Testing New End-User Information Systems: Theory and Results* (Massachusetts Institute of Technology, 1985), 291.
8   Davis, Fred D., "Perceived Usefulness, Perceived Ease of Use, and User Acceptance of Information Technology," *MIS Quarterly* 13, no. 3 (1989): 319–40.
9   Likert, Rensis, "A Technique for the Measurement of Attitudes," *Archives of Psychology* 140 (1932): 1-55.

10 Davis, Fred D., Richard P. Bagozzi, and Paul R. Warshaw, "User Acceptance of Computer Technology: A Comparison of Two Theoretical Models," *Management Science* 35, no. 8 (1989): 982–1003.

11 Cronbach, Lee J., "Coefficient Alpha and the Internal Structure of Tests," *Psychometrika* 16, no. 3 (1951): 297–334.

Case study

3D printed cartography in East Africa

Samuel Foulkes and Quentin Roa

Our organization, Clovernook Center for the Blind and Visually Impaired, in Cincinnati, Ohio, is the world's highest volume producer of braille materials. As a long-established nonprofit, Clovernook is committed to providing employment opportunities for people with visual impairments. Approximately 50 percent of our staff in the braille printing house are legally blind.

In 2020, Clovernook launched a Tactile Literacy Initiative, partnering with schools in Kenya, Uganda, Rwanda, and Tanzania. This program seeks to enrich classrooms with curriculum-aligned tactile materials. At first, the initiative focused on the provision of culturally relevant print/braille storybooks in local languages for early literacy classes, using 3D printed objects (mostly using PLA filament) to replicate items mentioned in the stories. These tactile replacements for illustrations in print materials help bond words to meanings. For instance, a story of an elephant chasing a rabbit, who then hides inside a pumpkin, is accompanied by models of an elephant, a rabbit, and a pumpkin with a lid, into which the rabbit can be placed.

Although these books and kits remain central to our program, we discovered through needs assessments that there was a significant need for stand-alone tactile materials to supplement science, math, and geography subjects. Models in existing databases were often not appropriate for educational purposes, let alone for students who are blind or low vision, so we secured funding to begin designing our own models using iterative design. We conducted testing in both the United States and East Africa to ensure that the 3D models we developed are optimized for classroom use.

Nairobi makerspace. Image courtesy of Clovernook.

The increasing accessibility and quality of 3D printing makes it a promising resource for education, compared with traditional tactile formats for people with visual impairments. With this in mind, we have decentralized 3D printing operations, opening a makerspace in Nairobi, Kenya, that is locally staffed and able to print materials for our educational partners. This allows for faster, less expensive, and more efficient testing and operations—and perhaps, most importantly, brings training tools and methods directly into the hands of the community that the program seeks to serve. In its first year, our Nairobi Makerspace has produced more than a thousand models for education in the region.

For geography-related materials, we decided to start from scratch when designing maps of regions, where we aimed to provide various types of information about a country that educators requested, including elevation, cities, hydrography, borders, land cover, roads, railroads, and airports.

With so many features requested, we decided to create a series of maps for each region, highlighting different aspects in each model. The most important step was to establish a common tactile language. Our first map was key in terms of testing and evaluating tactile symbology with end users so we could use universal guidelines for all future maps.

Our first 3D map design was of Kenya, incorporating elevation, cities, bodies of water, and borders. Our initial iteration measured 20 × 20 cm, the maximum size for a single piece on a typical 3D printing bed. Based on satellite data, the map used raised topography to represent elevation; dotted lines for rivers; lower-height, mass-dotted texture for bodies of water; and a hyphenated line for the equator. Raised circles represented cities, with rectangular labels for other areas of interest. Cities and areas of interest were labeled with abbreviated names and titles in braille, spelled out in a supplemental key. The design also included a scale guide and an inset box showing Kenya's location within the continent.

Once we had our initial design, we printed copies in the United States and Kenya and began testing with end users and educators in both locations. This led us to the next iteration of the model. The initial map was too small for the number of elements we included—a classic case of too much "tactile clutter." We increased the map's size to 36 × 36 cm, now printed in four pieces that interlock using a simple dovetail design. We also made changes to the water texture and addressed the quality of the 3D braille characters.

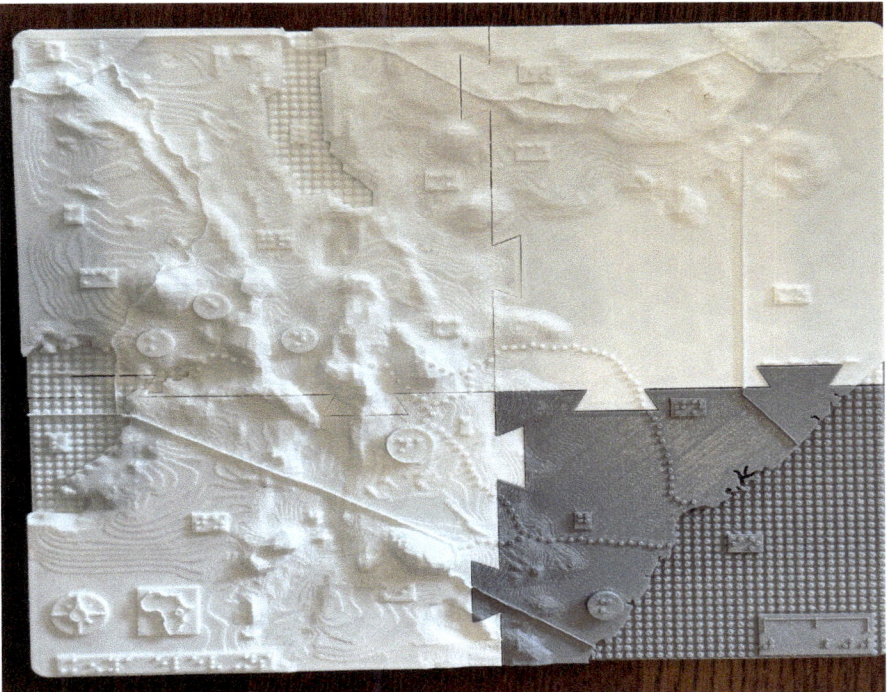

Second round of map of Kenya, with improvements made. Image courtesy of Clovernook.

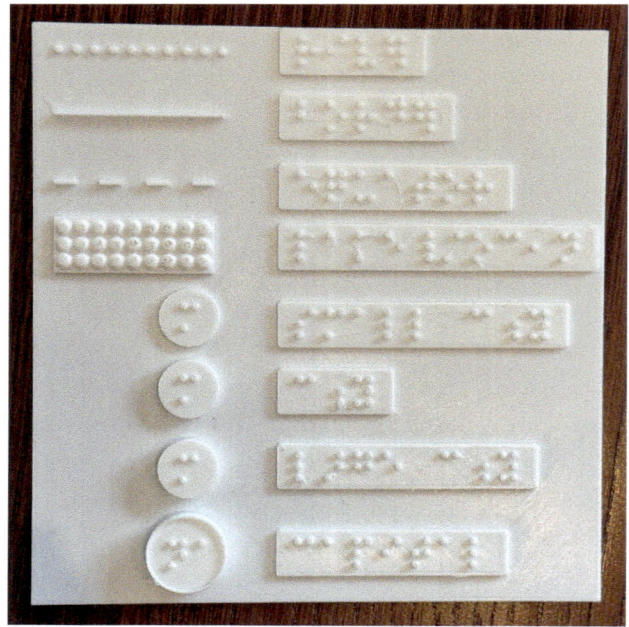

Kenya map partial key with names and labels in braille. Image courtesy of Clovernook.

Iterative testing and design work continues on our map series for Kenya, Uganda, Rwanda, and Tanzania. These maps will be available for download and will also be distributed to partner locations for classroom use. The final versions of the maps that we produce will be printed with multiple filament colors, creating high contrast for low-vision users. We anticipate that material costs for each map will be approximately $5 and that each map will have a much longer shelf life than traditional graphics.

Clovernook's project to provide these open-access 3D printed tactile maps aims to provide another pathway for educators and learners to advance learning goals in the classroom. The use of mainstream popular technologies and equipment means that these material types are less expensive; they also continue to be further improved, refined, and automated in ways that ensure further effective innovation for the education of students who are blind or low vision, at increasingly lower prices.

Case study

Optimized route planning for blind pedestrians

Sagi Dalyot and Achituv Cohen

Available route-planning apps, although useful for the general population, often overlook the specific accessibility and safety requirements of blind users, such as detailed information on sidewalk conditions, crossing safety, and potential obstacles. We sought to design a solution by developing a route-planning algorithm tailored to the spatial and environmental needs of blind pedestrians. For this project, we used open-source data from OpenStreetMap (OSM)—a collaborative mapping project that allows users to create and edit maps from an open geospatial database.

To define the spatial requirements for navigation by blind pedestrians, we collaborated with orientation and mobility (O&M) experts and conducted interviews with blind individuals. Through these consultations, the users identified key criteria for accessible maps, such as clear and safe pedestrian crossings, street type and surface conditions, detectable landmarks for orientation, and routes that minimize exposure to complex intersections and high pedestrian traffic areas. This information was then used to establish a set of spatial rules and parameters that could be encoded into the OSM-based algorithm, enhancing the database to better reflect real-world challenges faced by blind pedestrians.

The algorithm we developed was then rigorously tested in urban environments. Blind volunteers and O&M professionals were asked to evaluate the routes generated by the algorithm and compare them with those they would typically take. The routes created by the system were evaluated based on accessibility, safety, and adherence to the criteria outlined by blind individuals. The findings revealed that routes generated by the algorithm were both safer and more accessible than those offered by existing commercial navigation applications. For example, the generated routes prioritized streets with sidewalks while avoiding street furniture and surface problems and favored crossings equipped with assistive infrastructure, such as tactile paving and vibration signals. Moreover, the generated routes often closely aligned with the recommendations provided by O&M experts. We found that the system provided blind pedestrians with a heightened level of confidence in navigating urban spaces because the routes were tailored to avoid potential hazards and prioritize ease of navigation.

Several participants emphasized that although the optimized routes were generally longer than the shortest path, they did not perceive them as such. This was primarily because the optimized routes were simpler to navigate, avoiding complex areas, such as parks and squares. Overcrowded areas with, for example, musicians and vendors can disrupt blind pedestrians' sense of orientation. The routes also avoided potentially hazardous spots, such as intersections with multiple sidewalks, which can add cognitive overload. Participants noted that walking along main roads, as suggested by the route-planning solution, was beneficial because they could listen to sounds and sense the direction of moving vehicles for orientation.

After the field experiments, participants completed a questionnaire to evaluate the various spatial factors and criteria implemented in the route-planning solution.

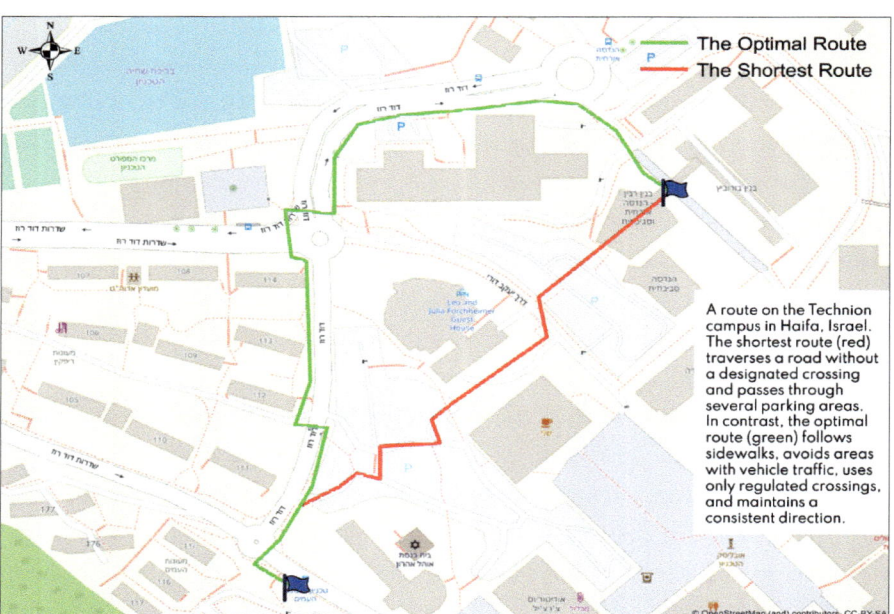

An example of optimal and shortest routes on the Technion campus in Israel. Image courtesy of Sagi Dalyot and Achituv Cohen.

These factors include information about street infrastructure, road furniture, commercial establishments, and natural features. Overall, each factor was relatively important for calculating the optimal route. However, if different weightings are needed for specific groups, such as white-cane users versus guide dog users, these can be customized to the user's needs (as, for example, user preferences in the navigation app) to derive optimal route planning. Differences also emerged among users from different geographic locations concerning the use of landmarks. For example, users in New York preferred the use of metro stations compared with a preference for bus stations in Haifa. Accordingly, the algorithm should allow for the selection of specific, prominent landmarks in a defined area to enhance the route's overall effectiveness.

Table. Average landmark importance for white-cane users versus guide dog users (higher scores mean greater importance)

Landmark	White Cane	Guide Dog
Traffic light	7	9
Bicycle rental/parking	5	1
Waste basket/bench	4	6
Restaurant/bakery	8	9
Supermarket	4	9
Cluster of trees	7	2
Bus/subway stop	8	10

This study shows how open-source platforms, such as OSM, can be adapted to serve specific user needs, proving that public geospatial data can be a great resource for developing accessibility-focused technology. That said, open-source data isn't perfect—it can sometimes lack accuracy or completeness. In areas with limited mapping information, the effectiveness of the algorithm might be impacted, highlighting the importance of regular updates and contributions from the community to improve OSM for these kinds of applications.

To sum up, our research offers a valuable step forward in creating accessible technology and better urban navigation for blind pedestrians. It emphasizes how combining accurate spatial data with a deeper understanding of users' needs can lead to truly inclusive navigation tools. With improvements in both data quality and algorithm performance, even greater progress can be made in providing safe and accessible city navigation for visually impaired individuals.

Personal story

Maps aren't just a fun gimmick

Parham Doustdar

Parham.

My wife and I have both been blind since birth, and we are good at getting around. We love to discover new places. We are highly skilled migrants who have been expats for almost a decade. We regularly visit other countries, and our first reaction when we're given the opportunity to do something new or go somewhere that we haven't been is "Yes!"

The technology we use to travel near and far, however, has its limitations. This can be life-threatening.

Once, when I was visiting a chiropractor for the first time in Amsterdam, the Netherlands, the navigation app on my phone gave me clear turn-by-turn directions, but it didn't tell me that there was a canal to my right. I took two wrong steps and fell two meters directly into the canal.

The scariest part was when I was in midair and didn't know how far I would plunge or what I would land on. What body part would I break? Would I be alive? Would I end up paralyzed? Was I falling onto a highway or into soft dirt? Fleetingly, I thought, "Life is fun! I'd be sad if it were over."

I survived that drop without any damage other than wet clothes. But the paramedics who pulled me out of the canal told me that if I had fallen five meters to the left or right, I would have at least broken my legs because the water was much shallower there.

Maps—and anything else that aids with mobility—aren't just a fun gimmick. They can help people get out of their comfort zones, participate in society, and achieve their dreams. At times, they can also save lives.

Conclusion

Vincent van Altena and Jakub Wabiński

This book explores the multifaceted world of tactile mapping, from the creation and use of maps to their impact on the lives of people with visual impairments. By inviting contributions from academics, practitioners, and experts by experience, we have tried to reflect the broad array of aspects in this field.

Although we adopted a multifaceted, multidisciplinary approach, we acknowledge that an important perspective is missing. Although several personal stories provide individual perspectives from people with visual impairments, and we both have experience in working with people with these challenges, there is no actual chapter with input from this community. Thus, we heartily invite our readership of people with visual impairments to reflect on the contents of this book and provide us with their insights and critiques. After all, our goal in publishing this book is that the target user group should benefit from it.

In our exploration of creating effective tactile representations, we gave due attention to factors such as the design of symbols, choice of production techniques, and cognitive processes at work while reading maps. But several authors also repeatedly emphasize the importance of user-oriented design and research, bringing the multiple perspectives and diverse needs of people with visual impairments to the fore.

The disability-rights mantra "nothing about us without us" pertains to the whole process of mapping. From the initial idea, content selection, map design and production to the choice of media, user involvement is mandatory and key to each step. Throughout this book, we advocate a holistic approach to tactile mapping.

Maps and perception

Throughout history, people with visual impairments have encountered various approaches to blindness and many obstacles, such as lack of facilities, training, and education. They have also had to overcome discriminatory societal attitudes, an evolution that is ongoing (see chapter 1). Access to maps and spatial information has been severely limited. Along with support for wayfinding and navigation, maps help us make sense of the world around us, be it on a local, regional, or global scale. Maps have fulfilled these functions for humanity since prehistoric times (see chapter 2).

However, applying and translating cartographic insights to tactile maps requires an understanding of the restrictions and challenges of touch (see chapter 3). Like the sense of sight, the sense of touch can mislead us (for example, with haptic illusions). Touch, however, poses even more challenges than sight because haptic resolution is much lower than visual resolution, requiring a more abstract, simplified presentation of details, as well as more distinctive features. The boundaries of tactile sensation must be respected in the design of maps.

Designing tactile maps

Developing a one-size-fits-all solution in tactile graphics is impossible. Individuals with visual impairments are diverse, not only in terms of their visual functioning and other sociodemographic characteristics that influence their level of independence but also in their access to facilities and eagerness to use them. Thus, unless we are developing custom-made materials tailored to the specific needs of an individual, tactile maps should remain legible to the widest possible audience. This requires not only tactile but also visual content developed in accordance with universal design principles.

The preparation of legible maps requires a balanced recipe that includes a translation of traditional "visual" cartography to its tangible counterpart (see chapter 4), an understanding of map design and cognition principles (see chapter 5), and generalization (see chapter 6).

The user and education

The human factor is paramount and entails an end-to-end involvement of the user. User-centered design and participatory research do not happen in a laboratory; taking the users' needs, wishes, preferences, and opportunities seriously throughout the process (see chapter 7) and interacting with users by experience is not only educational but also enriching for the designer. Every sighted designer, researcher, and academic would do well to act as a humble learner, seeking to understand what it means to be visually impaired and how to cope with daily challenges. Such an understanding should be checked repeatedly throughout the design process, and results can only be considered successful when the target group is willing to use them.

Even the most legible and beautiful map will remain useless if there are no readers to use it. Due in part to their scarcity, fragility, and relatively high cost, tactile maps are often stored in libraries or hidden in the back rooms of educational institutions (see the story of Johann-Ludwig Weissenburg in chapter 1). Mapmakers and issuers should ensure their adequate visibility and dissemination.

Also, education and training are key. Providing education not only acts as a gateway to otherwise inaccessible information but also trains potential map readers with the skills needed to explore maps and independently navigate the immediate environment and the larger world (see chapters 8 and 9). Part of this education and training involves creating awareness of tactile maps and simultaneously increasing their availability to potential users.

However, because of decreasing braille literacy, tactile reading (in general) may present a challenge to people with visual impairments, affecting their handling of tactile maps. Users often need to be trained in haptic sensations. On the other hand, this constraint might also open opportunities for presenting spatial information in new and different formats, including innovative multisensory outputs, combining touch, hearing, smell, and even sight.

Reliable output

New techniques, such as additive manufacturing and multisensory devices, along with progress in their design, offer inspiring perspectives (see chapter 10). These solutions, however, require proper evaluation by representatives of the target user groups. Only after positive feedback from potential users should they be put into production. That simply is usually not the case. Many assistive technologies dedicated to people with visual impairments never leave the prototype stage. It is not enough to gain more knowledge and insights from practitioners. Solutions must be put into the hands of the end user and meet their needs.

Yet, even if there is no one-size-fits all solution, we can still reuse insights, methods, and procedures as sources of inspiration (to say the least) or as templates that can be tailored toward our own specific requirements. This requires a methodical approach (see chapter 11) that ensures efficiency and allows creating solutions that are duplicable and generalizable, oftentimes thanks to actual user insights.

Reflections

Maps are tools

We don't want to exaggerate the importance of maps. They are tools among other tools to help us understand the world, and we need to choose the most adequate tool for the job. We can imagine contexts and settings where a tactile map may not be the most efficient medium of communication but could be better replaced by, for example, an audible or braille-printed sentence.

On the other hand, we shouldn't narrow the definition of cartography to paper or printed tactile maps. Cartographic spatial information can be communicated using a plethora of media, which we call "maps" as shorthand. As stated in chapter 2, "Maps play a critical role in modern societies, serving far beyond their traditional purpose of navigation. They are dynamic tools for analysis, planning, and understanding, with applications ranging from urban development and disaster response to environmental conservation and public health. … Maps remain indispensable tools for navigating not just spaces but also the interconnected challenges of modern life."

Rehabilitation

We are not living in a world of equal opportunities. Wealth, facilities, and access to health care are not equally distributed, either at a global scale or at a local scale. The same is true when we look at the distribution of eye diseases and the access to facilities to prevent, treat, or heal them wherever and whenever possible. Is there perhaps a special responsibility for the more fortunate to consider how we can share available resources? Instead of investing in further refinements of high-end technology, it could be wiser to see the ways in which as many people as possible can benefit from simple low-tech solutions.

At the same time, efforts should be made to rehabilitate people with visual impairments. Many of those adventitiously blind need support from people who will show them how to function in a completely new reality. In addition to professional orientation and mobility training, this requires everyday attention and involvement provided in a sensitive and respectful manner. Both neglect and over-involvement should be avoided. In this context, we would like to highlight the importance of institutions connecting people with visual impairments—both the official ones, such as national associations of the blind, and the less official groups on social media. It is our hope that this book will help raise awareness of the difficulties people with visual impairments face in their daily lives.

Wider application

Many people with visual impairments have residual vision and will benefit from recognizable visual cues. An implementation of primary colors also aids people with color vision deficiency and furthers communication between sighted users and those without sight. Combining multisensory output to communicate geographic information has the additional advantage of enriching someone's mental map with details not easily retrieved from a mono-sensory output (such as a tactile map). Interestingly, because tactile maps should be, in essence, highly generalized maps adapted for reading both by touch and by sight, they can also be successfully used by other groups of people with certain needs. Could, for instance, children or people who are easily overloaded with information or who suffer from attention disorders benefit from the sparse information communicated in tactile maps? We suggest further research into whether tactile maps could be appropriate to serve such information needs—research that also addresses, on a more conceptual level, the challenges of information selection and generalization.

Diversity and inclusion

Moving forward, we need a close collaboration between designers, educators, researchers, and people with visual impairments and an increased effort to promote a culture of inclusion. Besides being more effective and efficient, tailored toward user needs, such an approach also reflects the basic understanding that we are all part of humanity. We simply don't have the luxury of excluding anyone. On the contrary, being part of a society with diverse people requires thorough reflection to understand what it means to live together with those who have talents and challenges in life other than yours but also with whom you might heartily disagree. Such coexistence requires old virtues, such as empathy, charity, and willingness to step into someone else's shoes. In cartography as in life, we should recognize our interdependence by using each other's strengths to support each other's areas of need.

Looking ahead

We live in times of quickly developing technology. The rapid evolution of AI and large language models (LLMs) might allow effective automatic generalization, labeling, and generation of audio descriptions of tactile maps, reducing their costs and broadening the access to such materials. Recent developments in eye treatments, such as gene therapies and neuro-prosthetics, are offering new hope for individuals with visual impairments. Innovations such as retinal and cortical implants, as well as noninvasive stimulation methods, are enhancing functionality and could even restore vision in some people.

At the same time, new technologies require conscious and careful consideration, as technology-driven solutions can easily be carriers of exclusion (only being available, for instance, to the rich or tech savvy). This is not only a matter of economic advantage but also of uneven access to existing solutions, technology, and materials. Access of people with visual impairments to innovative solutions will differ considerably in urban and rural areas, for example, or in richer versus poorer nations.

No matter what the future brings and how accessible the new technologies will be, people with visual impairments must not only be motivated to use them but also have adequate education and training to use them properly. Likewise, professionals, practitioners, and researchers should use existing organizations to exchange knowledge and profit from others' insights. This begins in the specific domain of cartography, with research into participatory design, but is necessarily complemented by longitudinal, in-depth, multi- and interdisciplinary collaboration between national mapping agencies, associations, practitioners, researchers, stakeholders, and, most importantly, users.

We hope that this book will unite people with visual impairments with practitioners and stakeholders around the globe, inspiring them to develop legible and informative tactile maps. Such maps are not only aesthetically pleasing but have the potential to make an existential difference, providing the means for people with visual impairments to participate more fully in society. Some might say the production of tactile maps is just the proverbial "drop in the bucket," but for the person into whose hands the drop falls, they can be life changing. One such example is the story of Ellen, which you will find in the final pages of this book.

Personal story

I finally belong

Ellen Zieleman

Tactile maps don't restore my vision, but they do open and expand my world. Using these maps, I discovered, for instance, the expanse of forest in the Netherlands' National Park De Hoge Veluwe, which I have visited many times. And when I first felt a map of Europe, I realized that I only needed one fingertip to cover the Netherlands, but for Russia, I needed both hands.

With a tactile map of my surroundings, I can be more spontaneous. I can go out tomorrow instead of planning things weeks in advance. The maps let me know which direction to take when I step off the train, whether I'll pass through a shopping center or a park, and where it is safe to cross the street. They help me gain independence and make me feel like I belong.

With organizations around the world working together to produce tactile maps that perform well in local contexts, I hope that these kinds of resources become more accessible to everyone who needs them!

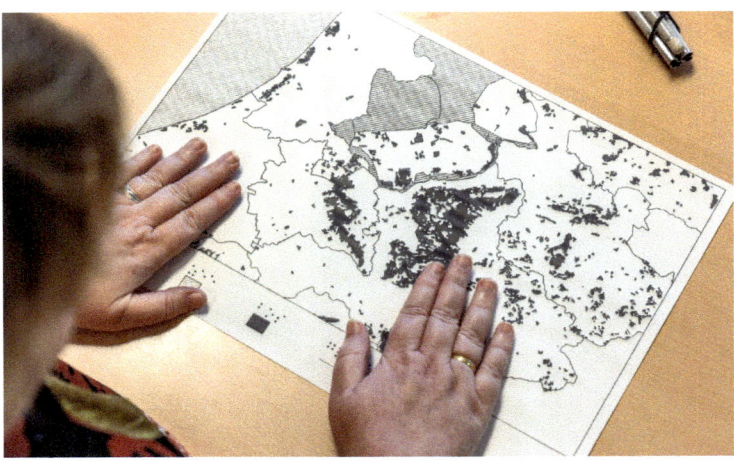

Ellen exploring a tactile map.

Acknowledgments

We extend our heartfelt thanks to all those who have contributed to this project. It would not have been possible to complete it without your dedication. Your personal stories and expertise and the efforts you made are priceless!

First, we would like to thank Ellen, Ran, Petr, Leydiane, Dorothy, Parham, Arend Jan, and Wojciech for sharing their stories. Their personal contributions opened our eyes to a world hidden from sight but filled with many lessons for us to learn.

In addition, we also want to express our gratitude to the practitioners and academics, who, despite the tight deadlines, contributed to the book. Georg, Astrid, William, Henrik, Ashna, Amy, Simon, Guillaume, Shirly, Rob, Merve, Zdeněk, Carla, Waldirene, Petr, Jolijn, Young-Hoon, Radek, Alena, Jan, Albina, Samuel, Quentin, Sagi, and Achituv: Working with such an interdisciplinary and international—global—team was new to us but very rewarding. Thank you for bearing with us!

Besides the people who are mentioned by name in this book, there have been several people active behind the scenes. Niels, thanks for your friendship and your tireless help in pitching and promoting the idea to Esri. The staff of Esri Press: Catherine, Stacy, Eric, Carolyn, and Victoria, and, most importantly, Jenny and Cici. Jenny, thanks for polishing our language throughout the book and sharing your experience in writing. Bouncing ideas off you was a lot of fun and very educating at the same time. Cici, thank you for your effort in phrasing the personal stories properly in English. These stories deserved beautiful language, and you provided them with such! Kadaster colleagues Bhavya and Marieke contributed by critically reviewing our (Jakub and Vincent's) contribution and steered us away from being either overly critical or not critical enough.

We extend our sincere gratitude to our employers, the Netherlands' Land Registry, Cadastre, and Mapping Agency and the Military University of Technology in Warsaw, for their generous support in providing the time and resources necessary to bring this book to fruition. We also acknowledge the invaluable contributions and encouragement from the International Cartographic Association (ICA) community. Many of the contributions to this book were written by ICA-affiliated authors.

In fact, the ICA brought us together—without it, we wouldn't have known each other and each other's research. The ICA champions inclusive cartography, aiming to ensure that everyone benefits from the power of maps and that cartography contributes to a more equitable world. Since 1984, a dedicated commission of the ICA has been directing its attention to tactile mapping. It is an honor to use this book to expand this body of knowledge and continue the work of previous commissions in the ICA's current Working Group on Inclusive Cartography.

For me, Vincent, special acknowledgments go to my beloved wife, Ruth. Not only for bearing—sometimes accepting and other times opposing—my mental absence while writing and editing this book, but more importantly keeping me aware that there's more to life than books, maps, and science. You taught me that living together and keeping an eye on one another is of the utmost importance.

As for me, Jakub, I would like to thank my wife, Emilia, from the bottom of my heart for her perseverance and understanding. I know how hard it was for you to take care of our two little children while I worked on the book after hours. Jerzy and Henryk will surely be as proud of their mother as I am of my wife!

Vincent van Altena, Zwolle, the Netherlands
Jakub Wabiński, Warsaw, Poland
March 2025

About Esri Press

Esri Press is an American book publisher and part of Esri, the global leader in geographic information system (GIS) software, location intelligence, and mapping. Since 1969, Esri has supported customers with geographic science and geospatial analytics, what we call The Science of Where. We take a geographic approach to problem-solving, brought to life by modern GIS technology, and are committed to using science and technology to build a sustainable world.

At Esri Press, our mission is to inform, inspire, and teach professionals, students, educators, and the public about GIS by developing print and digital publications. Our goal is to increase the adoption of ArcGIS and to support the vision and brand of Esri. We strive to be the leader in publishing great GIS books, and we are dedicated to improving the work and lives of our global community of users, authors, and colleagues.

Acquisitions
Stacy Krieg
Claudia Naber
Alycia Tornetta
Jenefer Shute

Product Engineering
Craig Carpenter
Maryam Mafuri

Editorial
Carolyn Schatz
Mark Henry
David Oberman

Production
Monica McGregor
Victoria Roberts

Sales & Marketing
Eric Kettunen
Sasha Gallardo
Beth Bauler

Contributors
Christian Harder
Matt Artz

Business
Catherine Ortiz
Jon Carter
Jason Childs

Related titles

The Power of Where

Jack Dangermond, Esri, and the GIS Community

9781589486065

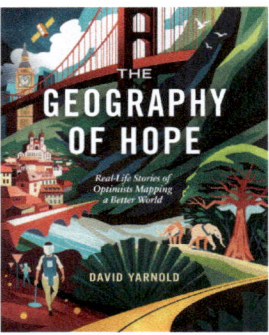

The Geography of Hope

David Yarnold

9781589487413

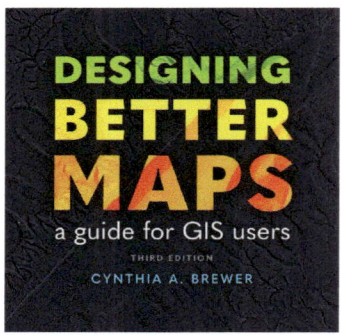

Designing Better Maps: A Guide for GIS Users, third edition

Cynthia A. Brewer

9781589487826

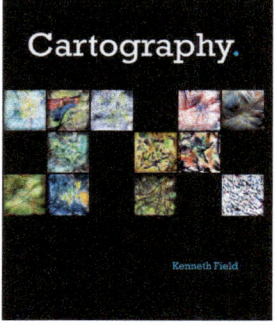

Cartography.

Kenneth Field

9781589484399

For more information about Esri Press books and resources, or to sign up for our newsletter, visit

esripress.com.